Probiotics

NIPA® GENX ELECTRONIC RESOURCES & SOLUTIONS P. LTD.
New Delhi-110 034

About the Author

Dr Gayathri Devaraja, Senior Professor, Department of Studies in Microbiology, Registrar (Administration) (Former), Registrar (Evaluation) (Former), Davangere University has a career spanning more than 28 years of Teaching, Research, Administration and Extension experience. She has published 100 plus peer reviewed research papers, undertaken Six major projects and four minor projects funded by UGC/DBT-DST/DAE-BRNS/VGST/KSCST, awarded Nine PhD degrees, received awards, fellowships and recognitions from various state, national and international governments/institutions. Her area of research includes Probiotics for healthcare system and Bioremediation. She is an awardee of most coveted prestigious Dr Kalpana Chawla Young Women Scientist by Government of Karnataka (GoK) with the jury Bharatha Rathna Prof CNR Rao, Best research publication by GoK, Three times "Best Project of the Year (by KSCST), Overseas fellowship to International vaccine Institute, Seoul, South Korea in addition to Young scientist and Best paper awards at National and international conferences, best international participant award by IVI, Seoul, S. Korea. As Dean of Science Faculty and Director, IQAC, has developed various programmes and policies for academic excellence.

Prof Gayathri has played a pivotal role in production of Monoclonal antibodies against *Aphanomyces invadans*, a causative agent of Epizootic ulcerative syndrome in Fish for the First time in the world and developed a rapid test, a detection kit with high sensitivity and specificity. With the financial support of VGST, KSTePs, GoK, Dr Gayathri has developed an exclusive Center for Innovation in Science, which specializes on Probiotic applications, bacteriocins for medical and industrial applications. Her major research focus is on development of Gluten reduced food for celiacs, Geriatric nutrition, alleviation of Lactose intolerance, Detoxification of aflatoxin and control of colorectal cancer using indigenous probiotic bacteria. The present book throws light on probiotics for students of UG, PG and research scientists and food technologists to promote natural food production using indigenous beneficial microbiome for healthy future.

Probiotics
Fundamentals and Beneficial Effects

Devaraja Gayathri
and
B.S. Rashmi

NIPA® GENX ELECTRONIC RESOURCES & SOLUTIONS P. LTD.
New Delhi-110 034

**NIPA® GENX ELECTRONIC
RESOURCES & SOLUTIONS P. LTD.**

101,103, Vikas Surya Plaza, CU Block
L.S.C. Market, Pitam Pura, New Delhi-110 034
Ph : +91-11-43860225, Mob.: +91 9717133558, 9540816132
E-mail: newindiapublishingagency@gmail.com
Website: www.nipaersources.com

Print ISBN: 978-93-58876-68-0

ebook ISBN: 978-93-58879-73-5

Composed and Designed by NIPA®.

Department of Science & Technology
Goverment of India

Technology Enabling Centre

(Established under Section 3 of UGC Act 1956)
Placed under Category 'A' by MHRD, GoI
Accredited with 'A+' Grade by NAAC

FAO Reference Centre for Antimicrobial resistance and Aquatic biosecurity

Foreword

The book on Probiotics is a useful resource for understanding the concepts related to probiotics and their effect on health. Their role in maintaining a healthy microbiome, fortifying the immune system, and mitigating various diseases is well presented. It will serve as a useful reference for beginners striving to understand them. This book presents the fundamentals of concepts related to probiotics, prebiotics, and synbiotics. Beginning with the taxonomy of two important probiotic groups of organisms, the lactobacilli and bifidobacteria, the book moves on to discuss the ecology of these organisms in the gastrointestinal tract, the production of exopolysaccharides and their role in controlling microbial infections in the gut and as potential therapeutic in infections of the urogenital tract. In keeping with advances seen in every field through molecular approaches and studies on the microbiome, the book discusses genome-wide analysis of gut bacteria. Apart from the health application, the book presents a glimpse of industrial applications of the enzymes produced by probiotics and antibiotics, such as Nisin from lactic acid bacteria, that find immense use as preservatives. The mechanism of action of probiotics ranges from the small antibiotic molecules that suppress pathogens to enzymes that aid in the degradation and synthesis of vitamins valuable for growth and antimicrobial effects through the production of hydrogen peroxide and acids by competitive exclusion of pathogens. The intimate interrelationship between diet, the immune system, and the microbiome is presented in the various chapters. Their role in "one health" based on their use in different sectors would add value to the existing topics covered in future editions. Probiotic effects have been attributed to the restoration of increased intestinal permeability, improvement of the immunological barrier functions, alleviation of the intestinal inflammatory response, and reduced generation of proinflammatory cytokines, reflecting their immunomodulatory functions. The book is a valuable reference material for students and scientists working in probiotics.

Nitte University Centre for Science Education and Research
Paneer Campus, Kotekar-Beeri Road, Deralakatte, Mangaluru-575 018, Karnataka
Tel: 0824-2204211 Mobile: +91 94484 79750
E-mail: indrani.karunasagar@nitte.edu.in, karuna8sagar@yahoo.com
Website: www.dstnutec.in

Preface

This book is a composition of various aspects on probiotics withholding innovative thoughts which is helpful to the students at graduation and post-graduation levels. It would be specifically useful to the young and senior scientists who intend to use and adopt the information into their academics and research. This book embodies various concepts and applications of probiotics. Academicians, researchers and students are encouraged to refer this book as it comprehends the basics of beneficial microorganisms, taxonomical aspects, enzymes, exopolysaccharide production, pre and synbiotics, next level identification strategies, bacteriocins; specially nicin and its application, in addition to upto date information on probiotics. This composition of the book evokes curiosity and generates interest for one to search for a novel probiotic species with suitable medical and environmental applications in a natural way. Hence this book is the right choice for microbiologists, food technologists, researchers, students and research scholars to be used as a ready resource material for the further advancement in their knowledge and applications at industrial level. We extend my sincere thanks to NIPA for their commitment and efforts in publishing this book. I place it on record with warm regards for the inputs rendered by my well wishers which provided a better shape to this book in a productive way.

Authors

Acknowledgement

This book is an outcome of 20 years of experience in probiotics and I thank my research students, collaborators, research grant agencies and mentors, who helped me in enriching knowledge with constant support. All my 12 research students and some post graduate students helped in reviewing the literature and created a strong database. My heartfelt thanks to all the Research Scholars of our Probiotic Laboratory particularly Dr Asha, Dr Shobha K Jayanna, Dr Somaraja P K, Dr Sunil Kumara S, Dr Swamy C T and Dr Vasudha M. My special thanks to Dr Rashmi B S, Research scholar, for strong commitment during her study, who contributed significantly in shaping this book. Many thanks to Dr Prashantkumar S Chakra, MS R Soundarya, Mrs Shwetha G S and Ms Pruthvi S M for contributing apt information at right time and their commitment in designing this book. Many of my MSc project students provided timely information in rewinding/improvising the Book.

I thank all the authors, reviewers and scientists whose works have been embodied in this book. I would like to thank UGC, DST-DBT, DAE-BRNS (Govt of India) and VGST (RFTT) and VGST-KSTePs (CISEE) for providing funding resources for the R & D works. Most of the information cited in the book are the derivatives of our published works in diverse peer reviewed journals such as Current Science, Experimental and Therapeutic Medicine, Current Microbiology, Journal of Basic Microbiology, Journal of applied Microbiology, Probiotics and Antimicrobial Proteins, Toxicon, J Mental Health Prevention, GenomeA, Biomedical Reports, Microbial Cell Factories, The Microbe, J Functional Nutrition and others. I wish to acknowledge the blessings of my mentors Dr K M Shankar, Professor of Aquaculture, College of Fisheries, Karnataka Veterinary, Animal and Fisheries Sciences University, Bidar, for instilling scientific spirit in me and Dr Indrani Karnunasagar and Dr Iddya Karunasagar, NUCCSER- KSHEMA, Deralakatte, Mangalore for always being caring and source of inspiration. My heartfelt thanks to my better half and best friend Dr Devaraja T N, Head and Chief Scientist, ICAR-TKVK, Davangere, for his unconditional motivation, support and providing scientific inputs time and again. My heartfelt thanks to our lovely daughters, Ms Abhijna Tanvi D and Ms Olavu Siri D, for their everlasting love and cooperation during the preparation of this book. My special thanks to my mother-in law and all my family members for the constant support in this enduring and memorable journey of scientific documentation.

Dr Devaraja Gayathri

Contents

Foreword .... *v*
Preface .... *vii*
Acknowledgement .... *ix*

1. **Fundamentals of Probiotics....... 1**
   Enhanced nutrient availability and absorption 4
   Production of bioactive compounds 5
   Maintenance of intestinal epithelial permeability and tight junctions in the host 5
   Criteria for the selection and characterisation of probiotics 7
   Role of the strain and dose in probiotic function 13
   Sources of Probiotics 13
   Groups of probiotic bacteria....... 14
   Key questions *17*
   Further reading *17*

2. **Prebiotics, Synbiotics & Nutraceuticals....... 19**
   Introduction 19
   Prebiotics and its characteristics 19
   Criteria for the selection of food ingredients as prebiotics 20
   Fermentation by intestinal microflora 20
   Selective stimulation of growth and/or activity of intestinal bacteria 21
   Types of prebiotics 21
   Some other important beneficial effects of prebiotics are as follows 24
   Nutraceuticals 25
   A dietary supplement 27

Key questions ... *30*
Further reading ... *30*

3. Taxonomy of Lactobacilli and Bifidobacteria ... 33
Introduction ... 33
Taxonomic classification of lactobacilli and bifidobacteria ... 33
Methods of taxonomic classification ... 34
Genomic organization of lactobacilli and bifidobacterium ... 37
Phylogenetic classification of lactobacilli ... 38
Phylogenetic classification of bifidobacterium ... 42
Key questions ... *45*
Further reading ... *46*

4. The Ecology of Lactobacilli in the Gastrointestinal Tract ... 49
Introduction ... 49
Interaction of lactobacilli with gut epithelial cells ... 50
Mechanism of metabolism in lactobacilli ... 51
Role of nutritional metabolism in genome organization ... 52
Reproduction in lactobacilli ... 52
Mechanism of adhesion of lactobacilli to the gastrointestinal epithelial cells ... 54
Colonization of lactobacilli in gut ... 56
Antagonistic, Antimicrobial and immunological properties of lactobacilli ... 57
Key questions ... *62*
Further reading ... *63*

5. Exopolysaccharide Production by Intestinal Lactobacilli ... 65
Introduction ... 65
Characteristics of the EPS ... 66
Biosynthesis of EPS and its Types ... 66
Homo exopolysaccharides (HoPS) ... 66
Hetero exopolysaccharides ... 68
Outline of biosynthesis of HePS. ... 70
Key questions ... *71*
Further readings ... *71*

6. Uses of Probiotics in Controlling Various Types of Microbial Infections; Bacterial Therapeutics 73

Introduction 73

Probiotics for infectious and antibiotic associated diarrheal diseases 73

Role of probiotics in inhibiting rotaviral diseases 74

Helicobacter pylori infections 75

Probiotics for preventing colon cancer 76

Production of protective metabolites 78

Bacterial therapeutics for the treatment and prevention of Urogenital infections 78

Management of Irritable Bowel diseases through probiotic and prebiotic interventions: 80

Lactose intolerance 84

Celiac disease 85

Key questions 87

Further readings 87

7. Genome Sequences: Approaches to Genome Wide Analysis of Gut Bacteria 89

Diversity of gut microbial community 89

Methods of counting viable and non-viable bacteria in gut 89

Techniques for enumeration of cultivable and non-cultivable microorganisms 90

Genome based insights for identification of microorganisms 90

Types of genome sequencing methods and its applications 93

Key questions 98

Further reading 98

8. Probiotics Enzymes, Industrial Applications 99

Introduction 99

Probiotics synthesized enzymes and its applications 99

Probiotics in aquaculture 101

Probiotics in poultry 104

Probiotics for farm animals 105

Probiotics for human 108
Key questions *111*
Further reading *111*

9. Diversity of Nisin Producing Lactic Acid Bacteria from Fermented Foods 113
Introduction 113
Applications of Lactic Acid Bacteria 115
Bacteriocin 116
Types of Bacteriocin 117
Bacteriocins from gram negative bacteria 117
Bacteriocins from Gram positive bacteria 118
Structure of Nisin 123
Mechanism of action of nisin 124
The mechanism of lantibiotic action is most extensively studied for nisin (Fig. 6). 125
Nisin variants 125
Nanotechnology and food integration 126
Applications of nisin 126
Infectious Diseases 126
Nisin as a cancer Therapeutic agent 127
Nisin, as modulator of gut microbiota 127
Antimicrobial Packaging Using Nisin 128
Key questions *128*
Further reading *129*

10. New Insights on the Role of Lactic Acid Bacteria (LAB) in Immune Modulation 131
Introduction 131
LAB host cross-talk and stress chaperons 133
Effect of LAB on the Immune Response 134
LAB adhesion to host tissues and immunomodulation 134
Key questions *135*
Write a short notes on *136*
Further reading *136*
Index 137

1

Fundamentals of Probiotics

The term probiotic means "for life" and it is currently used to refer bacteria that are associated with beneficial effects for humans and animals. The credit for first observation made on the positive role of some selected bacteria was attributed to Eli Metchnikoff, the Russian born Nobel Prize recipient, worked at the Pasteur Institute in the beginning of the last century. Metchnikoff suggested that - "the dependence of the intestinal microbes on the food makes it possible to adopt measures to modify the microflora in our bodies and to replace the harmful microbes by useful microbes". Interim Henry Tissier, a French paediatrician, observed that children with diarrhea in their stools had a low number of bacteria which are characterized by a peculiar Y shaped morphology. On the contrary, these "bifid" bacteria were abundant in healthy children. Based on these observations, Tissier suggested that "bifid" bacteria could be administered to patients with diarrhoea to help restore a healthy gut microflora. These two scientists were the first to give scientific intimation for the use of probiotics when even the term probiotics was not coined. Later in 1965, the term probiotics is coined by Lilly and Stillwell to describe substances produced by one microorganism, that stimulate the growth of another. Subsequently, Parker defined probiotics as organisms and substances that contribute to intestinal balance. Further, Fuller (1989) improved the Parker's definition of probiotics as "live microbial feed supplements which beneficially affect the host animal by improving its intestinal microbial balance". Havenaar and Huis in't Veld (1992) extended the probiotic definition with the description: "A viable mono or mixed culture of microorganisms which, applied to animal or man, beneficially affects the host by improving the properties of the indigenous microflora". At the end of the millennium, Salminen and his coworkers (1999) proposed the following definition: "probiotics are microbial cell preparations or components of microbial cells that have a beneficial effect on the health & well- being of host". Whereas FAO/WHO (2001) defined probiotics as "living bacteria that, when administered in adequate amounts, confer a health benefit on the host". This definition for probiotics is widely accepted.

Research works in the last decade showed that combination of probiotic therapy and conventional therapies helps to fight existing diseases and protect from future infections since probiotics have the potentiality to fight infections and also to modulate the immune system. The ability of probiotics to activate and enhance the host immunity is one of the beneficial aspects of probiotics on human health. But still, the mechanism of probiotic action has to be deciphered by continuous research in near future. However, existing knowledge from current research explains the mechanism of probiotic action as follows:

Competition for adhesion: Gut microflora comprise thousands of species. There will be continuous competition between the microorganisms of different species in order to establish their niche in the habitat. Therefore, adhesion to the epithelial cells is important to establish colonization, resisting peristaltic movement. Proiotic bacteria compete with the harmful microorganisms for the adhesion site on intestinal epithelial cells and prevent the binding of some of the pathogens and thus confer protection to the host against harmful microorganisms. Several factors have proven to influence the adhesion of probiotics. Bacterial concentration, buffer composition, growth medium and incubation time are significant *in vitro* conditions that affect adhesion. Whereas composition of normal intestinal microbiota, digestion and food matrix are the *in vivo* conditions that have an influential role on adhesion properties of probiotics. To consider, when probiotic species exhibit high affinity of probiotics towards receptors than pathogens or when they are high in their number, they have been proven to displace adhering pathogens. However, many studies failed to correlate *in vitro* adhesion and pathogen exclusion, questioning the role of adhesion property of probiotics in pathogen exclusion. These observations suggest that a more structured investigation of the relationship between adhesion and probiotic functionality is required.

Competition for nutrients: Nutrient requirement of individual commensals is one of the critical factor that determines the composition and distribution of microbiota in the intestine. One of the strategies used by probiotic bacteria to mount colonization resistance against pathogens is the preferential consumption of nutrients for the growth of competing pathogenic bacteria. Consumption of common limited nutrient sources by beneficial bacteria force competing pathogenic bacteria to starve and limit their growth rate. For example, commensal *E. coli* competes with Enterohemorrhagic *E. coli* (EHEC) for organic acids, amino acids, and other nutrients and confers protection to the host against EHEC infection, reducing its number.

Production of antimicrobial compounds: Probiotics produce an array of metabolites that are inhibitory to pathogenic bacteria. Organic acids such as acetic, propionate, butyrate, and formic acids lower the colonic pH and creates acidic environment which is unfavorable for the growth of most of the pathogens. For example, *Bifidobactrium* prevents the colonization of Pathogenic *E. coli* by acidifying its environment. Further, for most of the bacteria optimum pH is essential for secretion of proteins and other virulent factors. To mention, pH is a critical factor in the growth of *Bacillus cereus* and its enterotoxin secretion. In addition, butyrate downregulates the expression of several virulence genes including those encoding for Type 3 secretion system proteins in *Salmonella enteria* Serovar Enteriditis and Typhimurium. Further, probiotic bacteria suppress pathogen virulence by altering the intestinal conditions that are essential for virulence activity. For example, *Shigella flexeneri* requires ambient oxygen tensions for the secretion of its virulence factors. But the facultative anaerobes and microaerophilic bacteria consume residual oxygen in the intestine resulting in the incomplete expression of virulence factors. Furthermore, most of the Gram's positive bacteria produce short peptides called bacteriocins. Bacteriocins are ribosomally synthesized extracellularly released bioactive peptides or peptide complexes, having bactericidal or bacteristatic activity. But the producer cell exhibit specific immunity to the action of its own bacteriocin, producing specific immune proteins and cell wall bound transporter systems. Bacterial cell wall bound immune proteins bind bacteriocins and prevent them from entering producer cells. While, transporter system transports bacteriocin molecules that have been inserted into the bacterial membrane back to the surrounding medium and thus keeping the concentration of the bacteriocin in the membrane under critical level. These antagonistic properties of probiotic bacteria exclude pathogens and confer protection to the host by infections.

Modulation of immune system: Probiotics modulate host immunity by interacting with the Toll Like Receptors (TLR) on eukaryotic epithelial, endothelial and lymphoid cells. They stimulate the production of several immune compounds. They also produce antagonists of inflammatory cytokines and maintain a healthy balance between pro and anti inflammatory responses. Probiotics also induce apoptosis features a mechanism of programme cell death such as chromatin condensation, formation of apoptotic bodies DNA laddering, cell cycle arrest, accumulation of reactive oxygen species disruption of mitochondrial transmembrane potential, caspase activation and release of cytochrome C in the host. Further, probiotics promote mucosal barrier function and enhance innate immune responses. Bacterial signaling through MyD88 results in the release of sIgA by intestinal epithelial cells that binds to microbial antigens, neutralize pathogen activity and prevent infection. They also play an important role in the production of IL-1β which is essential for neutrophil recruitment to eradicate pathogens. Furthermore, interestingly ability of the probiotics to adhere to the intestinal mucosa has found to be associated with the modulation of immune system. Highly adhesive strains of probiotics proven to have greater effect on the immune system than less adhesive strains. Intestine is the body's largest immune organ and intimate contact with it could be the possible explanation for the role of adhesion ability in modulating immune system.

Enhanced nutrient availability and absorption

The food ingested by the host has direct effect on gut microbial composition that directly influences the pattern of biochemical reactions in the intestinal lumen. The carbohydrates, proteins and lipids ingested by the host serve as nutrient sources for the intestinal microorganisms. They produce an array of enzymes like amylases, proteases, lipases, phytase to hydrolyze the complex nutrient molecules to simpler ones. This increase the bioavailability of the ingested nutrients to host and the simple molecules improve the rate of nutrient absorption. Further consumed nutrients such as vitamins, amino acids, sugars, dietary fibres are assimilated and converted into other metabolites by gut bacteria. The product of these biochemical conversions such as short chain fatty acids (SCFAs), biogenic amines, amino acid derived metabolites such as GABA, serotonin, acetyl choline etc., play a key role in host physiology. Probiotics ferment indigestible dietary carbohydrates and produce SCFAs that provide metabolic energy and also involve in many biological process.

Futher, the nutrients components of the ingested food are not completely available for their absorption. For example, phytate in plant based foods such as legumes and cereals is an anhydrous storage form of more than 80% of the total phosphorous in cereals and legumes. Phytate has antinutrients behavior and has a potential for binding positively charged multivalent cations, proteins and amino acids in foods. Phytase, an enzyme that can breakdown the undigestible phytic acid and thus release digestible phosphorus, calcium and other nutrients. Probiotics with phytase activity hydrolyze phytate, releasing bound nutrients and makes them available for absorption. Thus, probiotics improves nutrient bioavailabilty and nutrient absorption in the host.

In addition, probiotics also proven to improve digestion in host and alleviate the immune burden associated with digestion. For example, wheat gluten is a complex protein rich in proline and glutamine amino acid residues. Human pancreatic, gastric and intestinal enzymes lack post prolyl activity and results in the partial digestion of gluten yielding smaller fragments that deposit on intestinal membrane and cause gluten intolerance. Further, Lactose intolerance, another food intolerance is the inability of adults and children to digest lactose, a sugar found in milk and to a lesser extent dairy products, causing side effects. It is due to a lactase deficiency, or hypolactasia. Probiotics hydrolyze lactose in the milk product or in the small intestine and alleviate the symptoms. Probiotics thus hydrolyzing such complex sugars and proteins, improve digestion in the host and confers protection against many such food mediated intolerances and allergies.

Production of bioactive compounds

Most of the vitamins and essential amino acids for our growth and development are being synthesized and supplemented by the gut microorganisms. The vitamins constitute anti oxidant system in host and confer antioxidant property. In addition, they also act as coenzymes in several metabolic pathways. Further, gut bacteria produce many metabolites that structurally and functionally mimic their eukaryotic host metabolites. They also had known to produce several neurotransmitters like serotonin, GABA, acetylcholine, tyrosine, tryptophan, biotin, catecholamines, vitamin K, *p*-hydroxyphenylacetic acid, *m*-hydrophenylacetic acid, hydroxycinnamic acid, phenyl valeric acid, *p*-aminobenzoic acid, indoxyl sulfate, indoleacetic acid, indolecarboxylic acid, indoleacetaldehyde, 6-hydroxymelatonin sulfate, putrescine, spermidine, spermine, taurine, cadaverine, and other peptides. Lipid metabolism by intestinal bacteria results in the generation of an array of fatty acid species. Species of lactic acid bacterial genera *Lactobacillus* and *Bifidobacterium* are known to produce conjugated linoleic, linolenic poly unsaturated fatty acids and the short chain fatty acids namely acetic, butyric and propionic acids; which are essential for cognitive function, development of brain and also known to alleviate the burden of inflammation. Compared to acetate, butyrate and propionate exhibit high anti inflammatory property. While acetate serve as primary substrate for the synthesis of cholesterol- a major constituent of human brain which is required for synapse and dendrite formation. Whereas Poly unsaturated fatty acids (PUFA) particularly ω-3 fatty acids also play an important role in cognitive function, neurogenesis, neurotransmission and confers protection from oxidative stress.

Maintenance of intestinal epithelial permeability and tight junctions in the host

Probiotics promote intestinal barrier integrity by the modulation of tight junction proteins. Probiotic bacteria were showed to upregulate the expression of TJ proteins particularly occludin and zonula occludins I, downregulating the expression of pore forming protein claudin-2 suggesting that probiotics could maintain intestinal permeability and tight junction integrity at transcriptional level directly. In addition, probiotics also showed to modulate various protein kinase pathways, enhancing phosphorylation of TJ proteins which can either elevate TJ formation and normal barrier function or facilitate the redistribution of TJ protein and complex destabilization. However, whether probiotic bacterial cellular component or their metabolites are responsible for the strengthening of TJ and normal functioning of intestinal barrier permeability have not yet been clearly understood.

In addition, probiotics particularly species of Lactic acid Bacteria have proven to take part in anti carcinogenic action, anti inflammatory effects, biotransformation of isoflavones phytoestrogen to improve post menopausal symptoms, bioconversion of bioactive peptides for antihypertension and also in alleviating symptoms of lactose intolerance. Probiotic bacteria can reduce the risk of occurrence of bacterial intestinal disorders and prevent diarrhea. Moreover, they are also characterized by anticholesterol activities. They reduce the cholesterol by producing lipase and also by assimilating fat in the body.

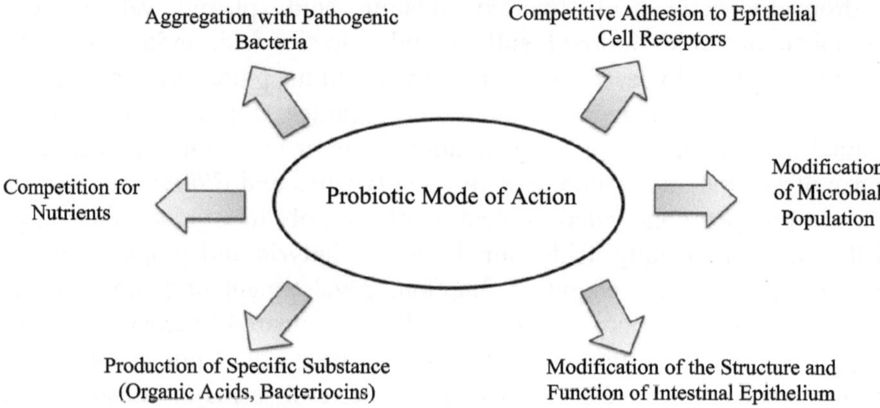

Fig. 1. Mechanism of action of Probiotics

Apart from this, successful probiotic bacteria should be able to survive gastric conditions and colonize the intestine, at least temporarily, by adhering to the intestinal epithelium. Such microorganisms appear to be promising candidate for probiotics. However, these beneficial characters are strain specific and more research is required to find out which strain works best for what conditions. The list of beneficial effects attributed to probiotics bacteria is extensive, in addition to alleviation of lactose-intolerance symptoms, gluten intolerance, serum cholesterol reduction, alleviating constipation, prevention of drug-induced colitis, while they also demonstrate efficacy in a number of conditions including ulcerative colitis, pouchitis, radiation colitis, atopic eczema, diarrhea and colon cancer. Successive chapters explain the role of probiotics in treating several diseases, while in this chapter the essentials of probiotics that are necessary to understand the basic concept of probiotics are discussed.

Criteria for the selection and characterisation of probiotics

Not all microbial isolates which are proven beneficial can be used as probiotics. There are certain criteria to screen the ideal and potent probiotic isolates for human as well as animal use. Earlier research yielded very poor results and consequently received very little attention. However, since 1990 probiotic research has gained much attention and come to the forefront of scientific research swiftly with significant progress being made in the development of clinically operative probiotic strains. Now a day, in market a wide range of probiotic products are available. For animals probiotics are being used as feed additives, to prepare silage and also to replace the use of antibiotics. In poultry and aquaculture to improve the yield and also to eliminate the use of antibiotics. Probiotics are available as health drinks, additives in various foods, extensively in diet as nutraceutical for people with nutrition related diseases and disorders, in infant formula, and also for various medicinal applications as therapeutic agents. Since the beneficial effect of a probiotic isolate is strain and host specific, therefore selection of the isolate varies with the host. In this regard care should be taken while selecting the probiotic isolate for a particular host such that a maximum beneficial effect can be obtained without any side effects.

In last decade, interest in the field of probiotic research has been increased vigorously and following factors may be the impetus for it.

- Continuous criticism and discussion of the research outcome using well documented probiotic organisms resulted in the exciting scientific and clinical findings.
- Publication of these exciting findings from the probiotics in contrast to the harmful effects of synthetic drugs increased the awareness about the limitations and side effects of the pharmaceutical agents.
- Subsequently, all these factors lead to the increased consumer demand for natural products. Increased awareness and consumer demand fetched the attention of research institutions and pharmaceutical companies on probiotics.

Although the concept of probiotics is not new, emergence of commercial products has refocused the attention on their applications and safety. A regulatory guideline for Probiotic foods is very important. Otherwise there would always be a chance for ersatz, spurious and ineffective products in the market. In this regard, to consider any microbial isolate as probiotic, it has to meet several selection criteria and possess a number of inherent physico-chemical characteristics. In 2001, FAO/WHO outlined the essential requirements for an isolate to be claimed as probiotic. In addition, several countries also developed their own guidelines for probiotic marketing in their nations and in most of the cases that would require probiotic strains to be backed by clinical trials preferably conducted in local population. Now a day there has been an increased influx of probiotic products in Indian market also. Therefore, as an initiative, Indian Council of Medical Research (ICMR) along with Department of Biotechnology (DBT), Govt. of India has constituted a task force that considered the guidelines available at different parts of the world and developed the guidelines for regulation of probiotic products in India. Scope of the guidelines is that it mainly deals with the use of probiotics in foods and it provides essential requirements for the assessment of safety, efficacy, and health claims of probiotic strains. Further, it also takes charge of labeling probiotic products. The key points mentioned in the FAO/WHO and ICMR-DBT guidelines are discussed hereafter in this chapter.

FAO/WHO and ICMR-DBT reports strictly recommend identifying genus, species and strains by both phenotypic as well as genotypic methods as a first step in the evaluation of candidate probiotic cultures. Since probiotic effects are strain specific, strain identity has considered as an important step in linking a strain to specific health effect and further it also enables accurate surveillance and epidemiological studies.Therefore it is relevant to characterize the isolated microorganism at strain level. For speciation of bacteria, widely accepted, most current and valid methodology must be followed. PCR based techniques, 16S rDNA sequencing and DNA finger printing techniques like pulse field gel electrophoresis and ribotyping are the current molecular techniques used for bacterial identification. DNA-DNA hybridization method is considered as reference method in assigning a strain to a species for which it belongs. But the technique is time consuming and requires large collection of reference strains which is difficult and beyond the resources for many laboratories. Therefore, 16S rDNA sequencing has been suggested as a substitute for DNA-DNA hybridization technique. Further, 16S rDNA sequencing genotypic technique is recommended to combine with phenotypic tests for confirmation. Fermentation patterns obtained for wide range of sugars and end products of glucose utilization are considered as key phenotypes for identification purposes.

Further, reproducible genetic methods or unique phenotypic trait has to be used for strain typing. Pulse Field Gel Electrophoresis (PFGE) is considered as gold standard for strain typing. Alternatively, Randomly Amplified

Polymorphic DNA (RAPD) can also be used, but it is comparatively less reproducible. In addition, screening for the presence of extrachromosomal genetic elements such as plasmids and their identification can also contribute for strain typing and characterization. After identification, nomenclature of the bacteria must comply with the current, scientifically recognized names as per the International Committee on Systematics of Prokaryotes (ICPS). Further, identified and nomenclature strains must be deposited in an internationally recognized culture collection/repository. Culture collection centers play a crucial role in the conservation and sustainable use of deposited cultures for various applications. Their main task is to collect, preserve and make microbial strains accessible to public for biotechnological applications, teaching and research purposes. Thus microbial culture collection centers contribute to the wealth of knowledge and also lead to the discovery of new taxa. Moreover, conservation of microbial strains constitutes part of a country's heritage.

After identification, further evaluation by *in vitro* tests with standard methodology has been recommended for screening of presumed probiotic strains. Resistance to gastric acidity, bile acid resistance, antimicrobial activity against potentially pathogenic bacteria (acid and bacteriocin production), ability to reduce pathogen adhesion to surfaces, bile salt hydrolase activity, adherence to mucus and/or human epithelial cells and cell lines, resistance to spermicides (applicable to probiotics for vaginal use) are currently practiced *in vitro* tests. These tests mimic hostile gut environment under *in vitro* conditions and are useful in understanding the mechanism of probiotic action and enhance knowledge about the strains. To explain further, viability and persistence in the gastro intestinal tract, immune modulation, and antagonistic properties are considered as the major functional characteristics of probiotics. In order to exert their functional properties, probiotics need to be delivered to the host in an active and viable form. Viability and activity of probiotics in the products are considered as obligatory parameters to achieve numerous health benefits. Most importantly the potential probiotic strain has to surmount the stress in gastric environment, resisting the effect of bile salts with the help of bile salt hydrolase and adherence to the host epithelial cells by exhibiting high affinity towards receptors for binding and with standing the flux of intestinal content. Following oral administration, in order to survive and proliferate in the intestine, probiotics have to tolerate several host environmental hurdles such as low pH in the stomach, toxic effect of bile in the proximal region of the small intestine, reduced water activity, and also elevated osmolarity particularly in the intestine. In addition, it has to tolerate the pH variations in the different regions of digestive tract that constitutes various organs from mouth to anus. During the initial process of digestion, ingesting food baths in saliva of the mouth with a pH 6.5 to 7.5. Mean while the probiotics in the food also come in contact with the saliva and survive the salivary pH and descends down the digestive tract entering the upper stomach with a pH range between 4 and 6.5. In the upper stomach, probiotics that resist the drop in pH from 6.5 of saliva to 4.0 enters the lower stomach characterized by high acidic environment with a pH below 4. Acid tolerance was thought to achieve by maintaining a constant gradient between cytoplasmic and extracellular pH by F_0F_1-ATPase, changes in the cell membrane and regulatory mechanisms, alterations in different metabolic pathways, and amino acid decarboxylation. Amongst F_0F_1-ATPase in the acid tolerant bacteria has significant role in conferring acid tolerance by ATP hydrolysis and proton translocation that generates a proton motive force *via* proton expulsion. As a consequence, the enzyme can increase the intracellular pH in the presence of extracellular low pH environment. Once the organism enters acidic environment, expression and synthesis of F_0F_1-ATPase subunits occur and regulation of which appears to occur at transcriptional level. Thus, surviving extreme pH of stomach is the crucial stage of their journey as probiotics in the gastrointestinal tract of the host. In stomach only their number reduces greatly and only those survived enter the duodenum with slightly raised pH range 7 to 8.5. In duodenum, they start to increase their number and continue their journey towards intestine. Thus facing all challenges once reaching gut, they must be able to colonize in the presence of pH range 4 to 7, proliferate and exert beneficial effect on the host. *In vitro* survivability of the isolate in low pH and high levels of bile allows the probiotics to survive through gastric transient in the stomach and helps to reach intestine and thus balancing gut flora. Several studies on improving the probiotic potentiality of isolates achieved this by subjecting the bacteria to stress *in vitro*. However, these tests are not fully adequate to predict the functionality of probiotic microorganisms in the human body and are also not sufficient for describing them as probiotic. Moreover, the cultures evaluated on these tests must be subjected to preclinical studies using appropriate animal models and then have to confirm efficacy in more than one human trail.

Although probiotics have been considered safe, theoretically they may be responsible for following four types of side effects namely: systemic infections, deleterious metabolic activities, excessive immune stimulation in susceptible individuals and gene transfer. Notably, correlation between the consumption of probiotics and these documented side effects are very few and all occurred in patients with underlying medical conditions. Further, provided importance for assuring safety, it has been recommended that probiotic strains should be characterized at a minimum with the following tests:

a) Determination of antibiotic resistance patterns. The putative probiotic strain must assure that it is not at significant risk with regard to transferable antibiotic resistance genes.
b) Assessment of certain metabolic activities that affects host physiology (for ex: D-lactate production, bile salt deconjugation).
c) Assessment of side-effects during human studies.
d) Epidemiological surveillance of adverse incidents in consumers (post-market).
e) If the strain under evaluation belongs to a species that is a known mammalian toxin producer, it must be tested for toxin production. One possible scheme for testing toxin production has been recommended by the EU Scientific Committee on Animal Nutrition (SCAN, 2000) and if it also belongs to a species with known hemolytic potential, determination of hemolytic activity is also required.

> Placebo- a medicine or procedure prescribed for the psychological benefit to the patient rather than for any physiological effect. Usually in the field of biomedicine research, substance that has no therapeutic effect used as a control in testing new drugs.

Further, many of the times strains which are proven efficient *in vitro* may fails to prove *in vivo* and may harmful to the host. Therefore, the isolates evaluated as probiotics based on *in vitro* tests must be subjected to preclinical validation using suitable animal models for safety and efficacy prior to human trials. In addition, probiotics must be tested for an effect on a variety of clinical conditions. Standard methods for clinical evaluations are consisted of Phase 1 (safety), Phase 2 (efficacy), Phase 3 (effectiveness) and Phase 4 (surveillance). Phase 1 studies focus on safety as discussed above regarding theoretical side effects of probiotics including systemic infections, deleterious metabolic effects, excessive immune activity and gene transfer. In case of phase 2 studies, randomized double blind, placebo controlled (DBPC) human trials should be undertaken to establish efficacy of the probiotic product *in vivo* testing compared with placebo. The DBPC human studies enrolled with sufficient subjects are preferred. Because, statistical significance enables the investigator to conclude that the observed result in human volunteers is due to the errors while sampling or actually reflects the characteristics of the probiotics tested. In addition, the *in vivo* results or findings corroborated by more than one independent work at different research institutions are preferable. Satisfying the above discussed criteria do not mean that the strain could be used as a probiotic; demonstration of lack of infectivity in an immune compromised animal model that assure safety to the host is also required. Probiotics delivered in food are not tested in phase 3 unless it claims to alter a disease state and the claim should be made on sound scientific evidence in human studies. To consider, in case of probiotic evaluated for the prevention or treatment of a specific pathogen then a clinical study must be design to ascertain its effectiveness in preventing the infection (preventive study) or treating the disease (treatment study) upon exposure to the pathogen.

Significance of viability: Probiotic viability is of biological, technological and economical significance. In addition to host associated stresses, probiotics also forced to encounter technical stresses during processing, formulation, packing of the products that hampers their longevity. Despite of these technical stresses, probiotics must survive in sufficient numbers during their shelf life often at different storage conditions. Several studies have revealed that some probiotic products in the market have deficiencies in the viability of probiotic strains. In many of the cases, most promising and clinically relevant probiotic strains are often rendered as unusable due to their physiological and technological fragility. This may be due to storage conditions i.e. varying temperatures, manufacturing or food technologies setbacks such as inappropriate packaging materials that could affect probiotic stability through variations in oxygen permeability and moistness. One of the major reasons behind loss of viability of probiotics in products is oxygen toxicity. Most of the probiotic candidates are microaerophiles or facultative anaerobes. When they expose to oxygen in food products, they become susceptible for oxygen toxicity. To some extent this can be overcome by gradually increasing the concentration of oxygen through subsequent passaging of the cells and enhance their adaptability to it. Alternatively, the use of different packaging systems for food products with oxygen impermeability, the inclusion of oxygen scavenging agents like ascorbic acid, microencapsulation of probiotic cells are also of choice. Other important problems associated with probiotic viability are temperature and moisture content. Therefore, optimum temperature and humidity must be maintained in storage places. In addition, foods in which they are ultimately formulated, probiotics should not produce any off flavors or textures and must be in viable form only but not growing. Considering all these factors, generally high numbers have been suggested to compensate for the possible loss in the numbers of probiotic organisms during processing and storage, passage through the stomach and the intestine. It is therefore recommended that the minimum counts of probiotic bacteria be 10^6 CFU/g of the product at the expiry date.

Role of the strain and dose in probiotic function

Strains of the same probiotic species can be different, which has been demonstrated both *in vitro* and *in vivo* studies. therefore, clinical results from one study are applicable only to the strains being evaluated in that study. It is also possible that different strains may have the same effects. As research continues, species- or genus-specific attributes are likely to be found. Identification of genes or gene systems for specific attributes may make it feasible to predict *in vivo* function. Functional genomics may greatly aid this research, which is progressing at a rapid pace. Currently, however, research specific to the strain or strain combinations should be used to substantiate claims of physiological benefits. Considering this the expert committee working on probiotic guidelines recommend to specify the genus, species, and strain designations for all probiotic strains in a product mandatorily. Probiotics are generally available in the form of capsules, powder, tablets, liquid or incorporated in foods. Depending on the strains the dose of probiotics also varies (Table 1) and should be based on levels found to be efficacious in human studies. Further, one dose level cannot be assumed to be effective for all strains. For example, the efficacy of *Bifidobacterium infantis* 35264 has been documented at 10^8cfu/day, whereas the recommended dose of VSL#3 (VSL Pharmaceuticals) is 1.8 x 10^{12} cfu/day, a 4-log cycle difference.

Table 1. Some of the common probiotic strains and their recommended doses.

Probiotic strain	**Recommended daily dosage**	**Preparations**
Lactobacillus rhamnosus GG	10 billion CFUs	Capsules, Therapeutic yogurts and fermented milks
Lactobacillus sp./ Bifidobacterium sp.	100 million to 35 billion CFUs, depending on preparation	Capsules, Powder, Therapeutic yogurts and fermented milks
Saccharomyces boulardii	250 mg to 500 mg	Capsules
Bacillu ssp.	107 -108 spores/g	Powder

Sources of Probiotics

The most common sources of these friendly bacteria are yogurt, cultured buttermilk, and cheese. Cultured buttermilk is made with bacteria that produce lactic acid, which makes the product sour. Some cultured cheeses use bacteria to finish the maturation of the cheese. Another fermented dairy product is kefir. Traditionally yogurt has only one or two bacteria whereas kefir tends to have several friendly bacteria. Other foods that are either produced by bacterial fermentation or process are Japanese Miso, Tempeh, sauerkraut, beer, sourdough bread, chocolate, kimchi, olives, and pickles. Table 2 lists some of the known commercially available probiotic products with their specific applications.

Table 2. Commercially available probiotic products

Product name	Species	Company	Applications
Nesvita	LAB	Nestle, India	Gut Probiotic
Vizyl	LAB	Unichem, India	Gastrointestinal Health
ViBact	*Bacillus mesentricus, Clostridium butyricum, Streptococcus faecalis, Lactobacillus*	sporogenes	Lamb Performance
Improval	*Saccharomyces cerevasie, Lactobacillus sporogenes*	Cadila, India	Vaginal infections
Amul	*Lactobacillus acidophilus,*	Amul, India	Diabetics
Binifit	*Streptococcus faecalis, Clostridium butyricum, Bacillus mesentricus*	LAB Ranbaxy, India	Antibiotic associated Diarrhoea (AAD)
Actimel®	*Lactobacillus casei DN 014001*	Danone, France	Immune booster
Yakult®	*Lactobacillus casei*	Yakult Honsha Co., Ltd, Japan	Gastrointestinal health
Proviva	*Lactobacillus plantarum*	Probi, Lund	Gastrointestinal health
Gefilus	*Lactobacillus GG*	Valio Ltd, Finland	Gastrointestinal health
Bifilac	*Clostridium butyricum, Bacillus mesentricus, Lactobacillus sporogenes*	Tablets India Pvt Ltd, India	Irritable bowel syndrome, Diarrhoea
LC1	*Lactobacillus johnsonii*	Nestle	Gastrointestinal health
Biosporin®	*Bacillus subtilis, Bacillus licheniformis*	Biofarm, Dniepropetrovsk, Ukraine Garars, Russia	Pyo-speticpost operational complications, bacterial vaginitis,
Enterogermina®	*Bacillus clausii*	Sanofi Winthrop SpA, Milan, Italy	Bacteriotherapy and bacterioprphylaxis of gastrointestinal disorders
Mitomax®	*Pediococcus acidilactici, Saccharomyces boulardii*	Imagilin Technology, USA	Supports digestive system, Reduces stress, Reduces diarrhea, Reduces vomiting, Relieves constipation, Decreases body odor, Enhances immune response

Groups of probiotic bacteria

The most commonly used probiotic strains include the lactic acid bacteria (LAB). LAB are Gram-positive low G+C content, acid-tolerant and nonsporeforming rods or coccobacilli. Among LAB, *Lactobacillus* is most widely used probiotics. Lacobacilli are aero-tolerant or anaerobic LAB, and are nutritionally fastidious. They are strictly fermentative and can be divided into three groups based on fermentation characteristics: obligately homofermentative, facultatively heterofermentative and obligately heterofermentative. They have been used for centuries in food production processes (yogurt, cheese, pickles). More than 100 species of lactobacilli are naturally found in the digestive, urinary, and genital systems. Furher, *Lactobacillus* has been used for treating and preventing a wide variety of diseases and conditions (Table 2).

Some of the lactobacilli found in foods and supplements are *Lactobacillus acidophilus, L. blugaricus, L. rhamnosus, L. plantarium, L. reuteri, L. brevis, L. fermentum, L. caucasicus, L. lactis, L. helveticus, L. salivarius, L. casei, L. johnsonii and L. gasseri.* Amongst, *L. blugaricus* is commonly used in yogurt fermentation. In addition, these lactic acid producing probiotics also synthesize Vitamins D and K. *L. plantarum* produces lactolin an antibiotic and can also synthesize L-lysine, an anti-viral amino acid.

Another important group of LAB with potent probiotic characteristics is *Bifidobacteria*. There are approximately 30 species of *Bifidobacteria* that make up about 90% of the healthy bacteria in the colon. They appear in the intestinal tract within few days of birth, especially in breastfed infants. *Bifidobacteria* that are used as probiotics are *Bifodbacterium bifidum*, *B. lactis*, *B. longum*, *B.infantis*, *B. thermophilum*, and *B. pseudolongum*. Studies have shown that the administration of *Bifidobacteria* at appropriate dosage alleviated the burden of Irritable Bowel Syndrome, dental cavities, and also improved blood lipids. Most of the commercial probiotic products with *Bifidobacterium* claims for conferring gastrointestinal health.

Streptococcus thermophilus, Enterococcus faecium and Leuconostoc are other groups of well known probiotic bacteria. *Streptococcus thermophilus* produces enzyme lactase and found effective in treating lactose intolerance. *Enterococcus faecium,* is normally found in human intestinal tract, *E. faecium* SF68 and *E. faecium* M-74 are well studied species of this genus. Whereas *Leuconostoc* are generally ovoid cocci often forming chains, have been used extensively in food processing before the concept of probiotics throughout human history, and ingestion of foods containing live bacteria, dead bacteria, and metabolites of these microorganisms has been practiced for a long time. Along with other probiotic species *Leuconostoc* is more extensively used in sauerkraut fermentation. Other members of the LAB include *Lactococcus* and *Streptococcus* that are important components of the endogenous microbiota in the human ileum and jejunum with more moderate densities in the colon are also known to be used in probiotic dairy products.

Further, as these organisms are Gram positive, recombinant proteins can be secreted without getting trapped in the periplasm. This property makes LAB species attractive vehicles for food grade production of proteins and enzymes. *Lactococcus lactis* and other species of LAB known to present transiently in human gut is used for the delivery of cytokine, protein and DNA vaccines. Transient colonization provides an advantage of using it as a delivery vehicle. Otherwise sustained colonization may lead to the overstimulation of the targeted pathway. The possibility of delivering proteins directly to the mucosa has opened up new methods of therapeutic treatment. On the contrary, *Escherichia coli* Nissle 1917 (EcN) has been considered as the only Gram-negative probiotic bacteria. The army surgeon Alfred Nissle originally isolated this strain in 1917 from the feces of a soldier during the First World War who, in contrast to his comrades, did not develop infectious diarrhea during an outbreak of the highly contagious organism *Shigella*. His observation suggested that EcN might provide colonization resistance to mucosal pathogens. Since then the probiotic effect and biosafety of EcN has since been extensively shown in numerous trials and underlined by its long medical history in Central Europe as a microbial remedy. At present, EcN is contained in a drug called Mutaflor, which is used for the treatment of both infectious diarrheal diseases & Irritable Bowel Disease. Furthermore, EcN has been administered to neonates to prevent the colonization of their digestive tract by multidrug-resistant pathogens. In addition to bacteria, one of the non pathogenic species of yeast *Saccharomyces boulardii* has also proven as probiotic and found to be effective in preventing and treating antibiotic associated diarrhea and traveller's diarrhea. It has also been reported to prevent the reoccurrence of *Clostridium difficile*, to treat acne, and to reduce side effects of treatment for *Helicobacter pylori*.

Probiotics are microorganisms living in humans which may favorably influence human health and prevent some diseases. They have formed a vital part of Mediterranean and Middle Eastern diets for thousands of years, in the form of fermented milk and vegetable products such as yogurt, kefir, acidophilus milk, sourdough, sauerkrauts, olivesand pickles. The relatively low rates of chronic, age-related diseases in these regions might be attributed to enhanced consumption of probiotic fermented foods. Now, research is catching up with this traditional wisdom in the form of accelerated scientific investigations into the broad spectrum health benefits of probiotics. Although the mechanism of probiotic action has not yet fully understood, there are numerous attempts to use probiotics as supportive therapy, mainly in gastrointestinal tract diseases. Other possible applications of probiotics in medicine require further clinical studies, concerning the mechanism of activity, determination of indications for use and optimal therapeutic dose.

Key questions

1. Probiotics means:
   a) For life b) To kill
   c) To live d) None of the above
2. The term "Probiotics" was coined by:
   a) Lilly and Stillwell b) Parker
   c) Robertson d) Palade
3. Organic acids produced by probiotic bacteria:
   a) Lactic acid b) Acetic acid
   c) Propionic acid d) All of the above
4. The vitamin synthesized by gut bacteria is:
   a) Vitamin A b) Vitamin K
   c) Vitamin E d) Vitamin C
5. Antiviral amino acid synthesized by *L. plantarum* is-
   a) Serine b) Lysine
   c) Glycine d) None of the above
6. Which of the following statement is not correct with respect to probiotics?
   a) They have beneficial effects on host
   b) They are immunomodulating
   c) Their probiotic effect is strain specific
   d) Probiotic bacteria need not to survive in the gastric environment

Further reading

Ayyanna, R., Ankaiah, D., & Arul, V. (2018). Anti-inflammatory and antioxidant properties of probiotic bacterium Lactobacillus mucosae AN1 and Lactobacillus fermentum SNR1 in Wistar albino rats. Frontiers in microbiology, 9, 3063.

Brandelli, A. (Ed.). (2021). Probiotics: advanced food and health applications. Academic Press.

de Melo Pereira, G. V., de Oliveira Coelho, B., Júnior, A. I. M., Thomaz-Soccol, V., & Soccol, C. R. (2018). How to select a probiotic? A review and update of methods and criteria. Biotechnology advances, 36(8), 2060-2076.

Dimidi, E., Christodoulides, S., Scott, S. M., & Whelan, K. (2017). Mechanisms of action of probiotics and the gastrointestinal microbiota on gut motility and constipation. Advances in nutrition, 8(3), 484-494.

El-Enshasy, H. A., & Yang, S. T. (2021). Probiotics, the Natural Microbiota in Living Organisms. Probiotics, the Natural Microbiota in Living Organisms (1st ed.). CRC Press. https://doi.org/10.1201/9781351027540.

Halloran, K., & Underwood, M. A. (2019). Probiotic mechanisms of action. Early human development, 135, 58-65.

Kim, H., Fugaban, J. I. I., Holzapfel, W. H., & Todorov, S. D. (2022). Selection of beneficial bacterial strains with potential as oral probiotic candidates. Probiotics and Antimicrobial Proteins, 14(6), 1077-1093.

Kim, H., Kim, J. S., Kim, Y., Jeong, Y., Kim, J. E., Paek, N. S., & Kang, C. H. (2020). Antioxidant and probiotic properties of Lactobacilli and Bifidobacteria of human origins. Biotechnology and Bioprocess Engineering, 25(3), 421-430.

Latif, A., Shehzad, A., Niazi, S., Zahid, A., Ashraf, W., Iqbal, M. W., ... & Korma, S. A. (2023). Probiotics: Mechanism of action, health benefits and their application in food industries. Frontiers in microbiology, 14, 1216674.

Minj, J., Chandra, P., Paul, C., & Sharma, R. K. (2021). Bio-functional properties of probiotic Lactobacillus: current applications and research perspectives. Critical Reviews in Food Science and Nutrition, 61(13), 2207-2224.

Mishra, S., & Acharya, S. (2021). A brief overview on probiotics: the health friendly microbes. Biomedical and Pharmacology Journal, 14(4), 1869-1880.

Shi, L. H., Balakrishnan, K., Thiagarajah, K., Ismail, N. I. M., & Yin, O. S. (2016). Beneficial properties of probiotics. Tropical life sciences research, 27(2), 73.

Yousefi, B., Eslami, M., Ghasemian, A., Kokhaei, P., Salek Farrokhi, A., & Darabi, N. (2019). Probiotics importance and their immunomodulatory properties. Journal of cellular physiology, 234(6), 8008-8018.

2

Prebiotics, Synbiotics & Nutraceuticals

Introduction

Probiotics, prebiotics and synbiotics are interrelated terms. Probiotic is discussed in detail in previous chapter. In this chapter, prebiotics and its effect on gut along with probiotics are emphasized. Prebiotics can be defined as non-digestible food ingredients that stimulate the growth and/or activity of bacteria in the digestive system in ways claimed to be beneficial to health. They were first identified and named by Marcel Roberfroid in 1995. He defined prebiotic as a selectively fermented ingredient that allows specific changes, both in the composition and/or activity in the gastrointestinal microflora that confers benefits upon host well-being and health. Further, FAO defined prebiotic as "a non-viable food component that confers a health benefit on the host associated with modulation of the microbiota."Whereas the term synbiotics refers to a product that contains both probiotics and prebiotics in the form of synergism. A synbiotic has been defined as "a mixture of prebiotics and probiotics that beneficially affect the host by improving the survival and implantation of live microbial dietary supplements in the gastrointestinal tract, by selectively stimulating the growth and activating the metabolism of one or a limited number of health promoting bacteria, and thus improving host welfare". As described above, the combined use of prebiotics and probiotics is often described as synbiotic, but the United Nations Food & Agriculture Organization (FAO) recommends that the term "synbiotic" be used only if the net health benefit is synergistic.

Prebiotics and its characteristics

As prebiotics changes the resident bacteria either by altering their proportions or activities, they are considered as functional but not as drugs. Further, the ability of prebiotics to ferment specifically makes them more efficient and able to alter the intestinal microbial composition towards a more 'beneficial' community. Intestine harbors thousands of bacterial genera comprising both harmful as well as beneficial species. Further, the resident bacterial microbiota of the human intestine comprises approximately 95% of the total cells of the body and thus play a key role in the host nutrition, health and disease. Therefore, alteration of microbial composition towards more beneficial community must be achieved by stimulating potentially health promoting genera but not the harmful groups. The ultimate aim of supplementation of prebiotics in human diet is to maintain beneficial gut microbiota.

Criteria for the selection of food ingredients as prebiotics

For a food ingredient to be classified as a prebiotic it must fulfill the following criteria:

- It has to Resist gastric acidity
- Upon ingestion, it should neither be hydrolyzed, nor be absorbed in the upper part of the gastrointestinal tract.
- It should be selectively fermented by one or a limited number of potentially beneficial commensal bacteria. (e.g. bifidobacteria and lactobacilli).
- It must be able to alter the colonic microbiota towards a healthier composition by selectively stimulating the growth of bacteria that has potentially associated with health and well being. For example by increasing numbers of saccharolytic species while reducing putrefactive microorganisms.

To consider any food ingredient as prebiotic, it must strictly meet all of the above mentioned criteria. Further, researchers suggested several standard methodologies to evaluate prebiotic characteristics. Following discussion emphasis on the standard methods for the evaluation of prebiotic properties.

Non digestibility: To evaluate non digestibility i.e. resistance to gastric acidity, hydrolysis by gastric enzymes and absorption, *in vitro* methods such as determination of the ability to resist acidic conditions and enzymatic hydrolysis (salivary, pancreatic and small intestinal enzymes) have been followed. Whereas recovery of putative prebiotic in feces of germ free rats or antibiotic treated rats constitute *in vivo* methods. In case of human, recovery of undigested molecules in distal ileum and in feces has been considered as direct method while, the methods that assess an increase in the glycaemia or insulinemia constitute indirect methods.

Fermentation by intestinal microflora

Further, to study the anaerobic fermentation of carbohydrates by fecal bacterial population *in vitro*, batch and continuous culture fermentation systems are commonly used. In batch culture, vessels are inoculated with either pure cultures of selected bacterial species or sometimes preferably fecal slurry along with the sugar to be studied. Whereas to reproduce physical, anatomical, and nutritional characteristics of gastrointestinal regions, multi chambered continuous culture systems have been developed and were found useful in predicting the extent and site of prebiotic fermentation. For *in vivo* fermentation studies in rat, oral administration of prebiotic to be investigated through drinking water or drinks followed by the collection of fecal samples and gastrointestinal contents of anesthetized rats is widely accepted method. Furthermore, to study fermentation in humans both direct and indirect methods can be used. In the direct method, followed by an oral administration of fixed dose of putative prebiotic, feces will be measured for the presence of testing prebiotic. Whereas in indirect method, test of breathe air at regular intervals of time after an oral administration of testing prebiotic for the concentration of gases, particularly hydrogen will be carried out.

Selective stimulation of growth and/or activity of intestinal bacteria

To conclude whether the testing prebiotic selectively promotes the growth of beneficial intestinal bacteria or not, study of fecal microbial population on a range of selective agar medium before and after ingestion of the testing carbohydrate is useful. However, this conventional method fails to give data on non characterized and also non cultivable intestinal microorganisms. To surmount, molecular approaches which are independent of culture techniques and encompass the screening of entire microbial diversity such as fluorescence in situ hybridization (FISH), polymerase chain reaction (PCR), gradient gel electrophoresis, direct community analysis can be used. So far research in prebiotics suggested that only transgalactooligosaccharide and inulin are the two prebiotics that fully met all above disussed criteria.

The daily dose of the prebiotic is not a determinant of the prebiotic effect, which is mainly influenced by the number of *Bifidobacteria* per gram in feces before supplementation of the diet with prebiotic. The ingested prebiotic stimulates the growth of indigenous *Bifidobacteria*, and further larger that population, the larger is the number of new bacterial cells appearing in feces. In connection with this, a new concept of "prebiotic index" is proposed and defined as "the increase in the absolute number of *Bifidobacteria* expressed divided by the daily dose of prebiotic ingested.

Types of prebiotics

Based on the structural properties prebiotics can be grouped as following types:

- Short chain prebiotics
- Full or longer chain prebiotics and
- Full spectrum prebiotics.

Short-chain prebiotics contain 2-8 links per saccharide molecule and they ferment faster, example: fructo oligosaccharides. Short Chain-Fructooligosaccharides (SC-FOS) are a mixture of oligosaccharides consisting of glucose linked to fructose units, which are not digested in the human small intestine but are fermented in the right side of the ascending colon where they specifically promote the growth of *Bifidobacteria.*

Longer-chain prebiotics contain 9-64 links per saccharide and they tend to ferment more slowly nourishing bacteria predominantly in the descending colon.e.g. inulin.

Full-spectrum prebiotics provide the full range of molecular link lengths from 2-64 links per molecule, and nourish bacteria throughout the colon, e.g. Oligofructose-Enriched Inulin (OEI).

The majority of research done on prebiotics is based on full-spectrum prebiotics since they nourishes beneficial bacteria throughout the colon. The most common synbiotic combinations available include *Bifidobacteria* and fructooligosaccharides (FOS), *Lactobacillus* GG and inulins, & *Bifidobacteria* and *Lactobacilli* with FOS or inulins. Xylooligosaccharides (XOS) is also an emerging prebiotic found mainly in bamboo shoots, fruits, vegetables, milk and honey. Generally, XOS are mixtures of oligosaccharides formed by xylose residues linked through β-(1→4)-linkages. Based on the number of xylose residues (2 -10) in XOS they are named as xylobiose, xylotriose, xylotetrose and so on. But for most of the food applications xylobiose is preferred. Oral administration of XOS significantly increases the Bifidobacterial levels in healthy people. Therefore, daily intake of XOS results in the increase in relative ratio of *Bifidobacteria* to the total intestinal microflora.Galacto-oligosaccharides (GOS), also known as oligogalactosyllactose, oligogalactose, oligolactose or transgalactooligosaccharides (TOS), belong to the group of prebiotics. The composition of the galacto-oligosaccharide fraction varies in chain length and type of linkage between the monomer units. Galacto-oligosaccharides are produced through the enzymatic conversion of lactose.

Although, prebiotics and probiotics associated benefits are favorable, researchers are vigilant in making firm conclusions since benefits vary, depending on type and amount of pre- and probiotic consumed. Therefore, more human studies need to be done to understand their direct effect on health. Two Non digestible oligosaccharides namely transoligosaccharides (TOS) and inulin are amongst the most *in vitro and in vivo* studied prebiotics. TOS and inulin escape digestion in the upper gastrointestinal tract and reach the large intestine intact. In pure culture studies, all of the tested bifidobacteria, bacteroides, most of the lactobacilli and enterobactria, and some of the streptococci fermented inulin and TOS. For these sugars, bifidobacteria showed comparatively vigorous growth. In animal studies, these prebiotics improved bifidogenic effect with a significant decrease in the number of enterobacteria. All these observations projected them as ideal prebiotics for intestinal health. In addition, they have been reported to improve the calcium absorption in human. Further, significance of bifidogenic effect of breast milk in infants due to the presence of galactose as GOS has been well documented. The predominance of bifidobacteria in breast fed infants is thought to be due to their ability to utilize GOS in milk. Fortified cow's milk based infant formula with GOS alone or along with other oligosaccharides also proved to give the bifidogenic effect as human milk. Moreover, galacto oligosaccharides and long chain fructo oligosaccharides in infant formula showed to modulate the intestinal flora and immune system as human milk does. Therefore, GOS are mainly used in infant milk formula, follow-on formula, and infant foods. Further, health benefits of GOS mainly include protection against enteric infections, increased mineral absorption, immunomodulation for the prevention of allergies and gut inflammatory conditions, trophic effects of short chain fatty acids on the colonic epithelium, fecal bulking and reduced toxigenic microbial metabolism that further mitigate the risk factors for colon cancer.

Further, FOS and inulin increases the growth of lactic acid bacteria and promotes butyrate and lactate production which may alleviate the burden of intestinal inflammation and also nourishment of colonocytes.To support this study on healthy human volunteers with supplementation of oligo fructose decreased Bacteroides, Clostridia and Fusobacteria, while inulin supplementation reduced the Gram-positive cocci. However, both oligofructose and inulin significantly increased the fecal Bifidobacteria. Further, these saccharides benefit gastrointestinal health by providing short-chain fatty acids (SCFAs), stimulating the proliferation of beneficial bacteria and suppressing potential pathogenic organisms. Further, SCFAs affect colonic epithelial cell transport, energy transduction in colonocytes, growth, and cellular differentiation. These properties have important physiological implications, in addition to strengthening the mucosal defense barrier against invading organisms. After absorption into the bloodstream, SCFAs affect the hepatic control of lipid and carbohydrate metabolism and provide energy to different organs. Propionate hinders fatty acid and cholesterol synthesis while lactate lowers triglycerol fatty acids. Inulin containing diet enhances fecal butyrate. Stimulation of butyrate production is of interest in preventing ulcerative colitis, Crohn's disease and colon cancer. Thus, production of SCFAs is considered as one of the most important microbe mediated physiological functions in the gut.

In the previous chapter and present chapter, we have discussed the beneficial effects of both prebiotics and probiotics. Studies showed that administration of both the probiotics and prebiotics in combination as synbiotic is more effective than using only either of them. Synbiotics are better immunomodulators than the probiotics or prebiotics given separately. For example, synbiotic treatment prevented azoxymethane-induced suppression of NK-cell activity in Peyer's patches, an effect not observed in the individual pro- and prebiotic treatments. Combination of inulin and *B. longum* was more successful at decreasing azoxymethane induced aberrant crypt foci (ACF) than either treatment alone. Consumption of synbiotics containing prebiotics such as fructo-oligosaccharides or fructans, and probiotics, such as *Lactobacillus* or *Bifidobacteria* decreased experimentally induced colon cancer in animals. Short-chain fructo-oligosaccharides have also been reported to reduce colon tumours in mice, a genetic model mimicking human colon cancer. Administration of probiotics and prebiotics separately may encounter practical difficulties in reaching gut and exerting their beneficial effects. Probiotics in the gut are exposed to harsh environmental conditions which may affect their survival. On the other side, if prebiotic is given separately, then much benefit cannot be expected. Hence, encapsulated synbiotics are more effective than unencapsulated form. Encapsulation protects probiotics from the gastric transit and releases safely in the gut along with synbiotics where they interact immediately and benefit the host.

Some other important beneficial effects of prebiotics are as follows

Increased synthesis of vitamins: Prebiotics stimulate proliferation of vitamin synthesizing bacteria particularly B complex vitamins and eliminates vitamin deficiency in the host.

Improved calcium and other mineral absorption: SCFAs produced as a result of prebiotic fermentation acidifies gut environment and increase solubility of minerals in the gut and improves their absorption. In addition, prebiotics induce the expression of calcium binding proteins in the large intestine and facilitate calcium uptake by the enterocytes. They also enhance the bioavailability of phyto-estrogens.

Reduction in the risk of colorectal cancer: With the production of SCFAs, reducing inflammation, enhancing immune function and anti–tumorigenic activity, binding to toxins & carcinogens, mitigate the activity of enzymes generating carcinogens, prebiotics proven to reduce the risk of colorectal cancer.

Reduction in the risk of inflammatory bowel disorders: SCFAs nourishing colon cells found to be beneficial in Crohn's disease and further ameliorates ulcerative colitis through reducing the level of hydrogen sulfide gas by eliminating sulfate producing bacteria.

Regulation of blood cholesterol and antihypertensive effect: Prebiotics produced SCFAs are absorbed by portal vein in the large intestine and metabolized by the liver subsequently affecting various metabolic processes resulting in lowering of blood pressure. Further, lactate reduces the synthesis of triglycerol fatty acids while propionate hinders cholesterol synthesis. By binding effect also, prebiotics reduce cholesterol levels. In addition, prebiotics absorb phospholipids and fat in distal colon and enhance their excretion in feces. Reduced levels of total cholesterol regulate the receptors of LDL and thus enhance the clearance of LDL cholesterol. This overall cholesterol lowering effect further reduces the stiffness of large arteries and thus potentially reduces blood pressure

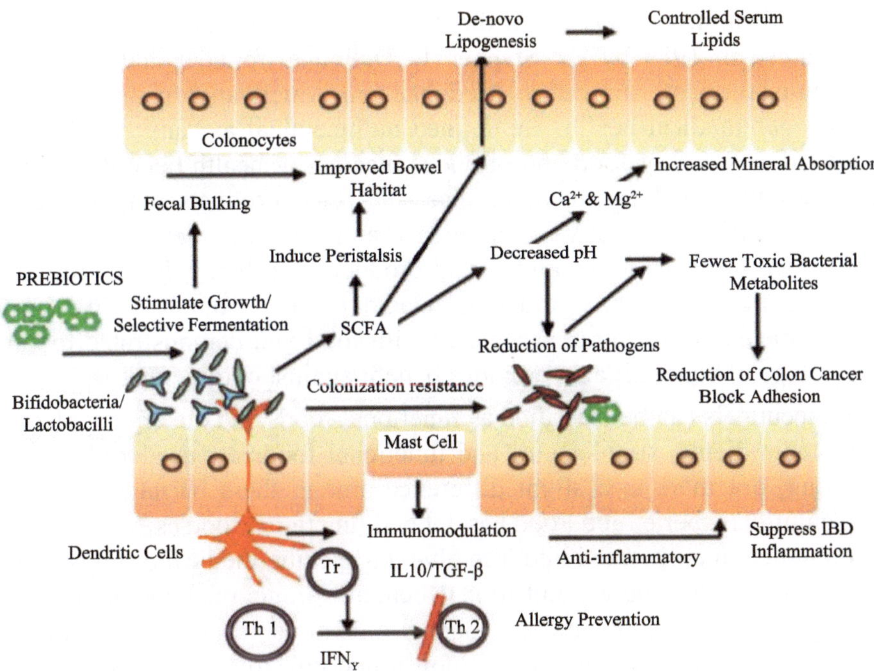

Fig. 1. Potential health benefits of prebiotics (source: Aachary and Prapulla, 2010).

Nutraceuticals

"Let your food be your medicine and your medicine be your food". This is a quote attributed to Hippocrates, the Father of Western Medicine which clearly establishes a linkage between diet and a person's overall health and well-being. Although it was made over 2,400 years ago, the fascinating term nutraceuticles made it is just as true now as it was back then.

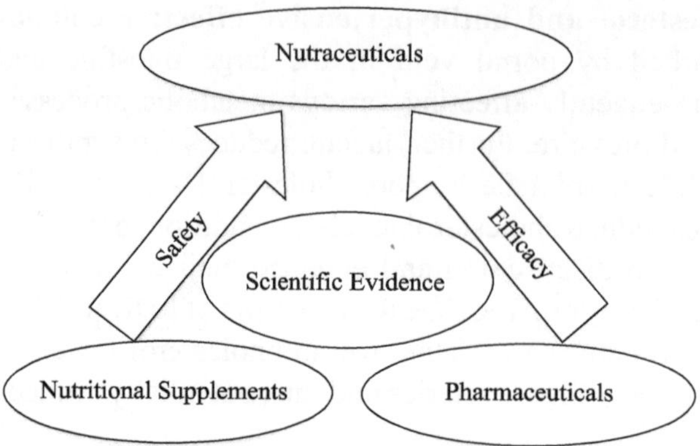

Nutraceutical, a portmanteau of the words "nutrition" and "pharmaceutical", was coined in 1989 by Dr. Stephen L. DeFelice, founder and chairman of the Foundation of Innovation Medicine, an American organization which encourages medical health. He defined nutraceutical as "any substance that is a food or a part of food and provides medical or health benefits, including the prevention and treatment of disease". Such products may range from isolated nutrients, dietary supplements and specific diets to genetically engineered designer foods and herbal products. Further, Ministry of Canada defines nutraceutical as "a product isolated or purified from the food, generally sold in medicinal form not associated with food and demonstrated to have a physiological benefit.It also provides benefit against chronic disease".

Neutraceuticals can be classified as functional foods and dietary supplements. Often both dietary supplements and functional foods are used with the same meaning, but there is a slight difference among them. Britain Ministry of Agriculture, Fisheries and Food has defined functional food as "a food that has a component incorporated into it to give it a specific medical or physiological benefit, other than purely nutritional benefit". Functional foods are similar in appearance to conventional foods. In contrast to conventional foods, functional foods, however, have demonstrated physiological benefits and can reduce the risk of chronic disease beyond basic nutritional functions, including maintenance of gut health. When food is being cooked or prepared using "scientific intelligence" with or without knowledge of how or why it is being used, the food is called "functional food". Thus, functional food provides the body with the required amount of vitamins, fats, proteins, carbohydrates, *etc*., needed for its healthy survival. Functional foods are designed to allow consumers to eat enriched foods close to their natural state, rather than by taking dietary supplements manufactured in liquid or capsule form. Functional foods have been either enriched or fortified, a process called nutrification which restores the nutrient content in a food back to similar levels from before the food was processed. Sometimes, additional complementary nutrients are added, such as vitamin D to milk (http://www.fao.org/ag/agn/agns/files/Functional_Foods_Report_Nov2007.pdf).

In the Japanese market, the functional foods are known as "Foods for Specified Health Use" (FOSHU). The functional foods comprise the following:

1. Conventional foods containing naturally occurring bioactive substances (e.g., dietary fiber).
2. Foods enriched with bioactive substances (e.g., probiotics, antioxidants).
3. Synthesized food ingredients introduced to traditional foods (e.g., prebiotics).

Whereas a dietary supplement is a product that contains nutrients derived from food products that are concentrated in liquid or capsule form. In the US, the Dietary Supplement Health and Education Act (DSHEA) of 1994 defined as 'A dietary supplement is a product taken by mouth that contains a "dietary ingredient" intended to supplement the diet'. The "dietary ingredients" in these products may include: vitamins, minerals, herbs or other botanicals, amino acids, and substances such as enzymes, organ tissues, glandulars, and metabolites. Dietary supplements can also be extracts or concentrates, and may be found in many forms such as tablets, capsules, softgels, gelcaps, liquids, or powders".

The DSHEA formally defined "dietary supplement" using several criteria.

A dietary supplement

- is a product (other than tobacco) that is intended to supplement the diet that bears or contains one or more of the following dietary ingredients- a vitamin, a mineral, a herb or other botanicals, amino acids or a dietary substance for use by man to supplement the diet by increasing the total daily intake or a concentrate, metabolite, constituent, extract, or combinations of these ingredients.
- is intended for ingestion in the form of pill, capsule, tablet or liquid.
- is not represented for use as a conventional food or as the sole item of a meal or diet.
- is labeled as a "dietary supplement". (http://vm.cfsan.fda.gov/~dms/dietsupp.html).

Considering the promising effects of nutraceuticals, they are considered in two ways:

- Potential nutraceuticals *and*
- Established nutraceuticals (Fig. 2).

A potential nutraceutical is one that holds a promise of a particular health or medical benefit; such a potential nutraceutical becomes an established nutraceutical only after there are sufficient clinical data to demonstrate such a benefit. A potential nutraceutical is one with promising particular health or medical benefit. To establish such a neutraceutical, it has to demonstrate its benefit in several clinical studies that are independent of each other. It is disappointing to note that the overwhelming majority of nutraceutical products are in the 'potential' category, waiting to become established. The food products used as nutraceutical are categorized as.

- Probiotic
- Prebiotic
- Dietary fiber
- Omega 3 fatty acid
- Antioxidant

Diverse health benefits extended by probiotics led to their application as functional foods. Several food supplements and nutraceutical products containing probiotics are developed and commercialized in several application forms like capsules, tablets, powders etc. Fermented food containing live cultures of probiotics are considered as fermented functional foods. Probiotics are commonly consumed as part of fermented foods such as dairy products like yogurt, kefir, kumis, cheese and fermented non dairy products like meat, soy yogurt, yosa, ogi, sauerkrauts, fruits, vegetables and malt beverages.

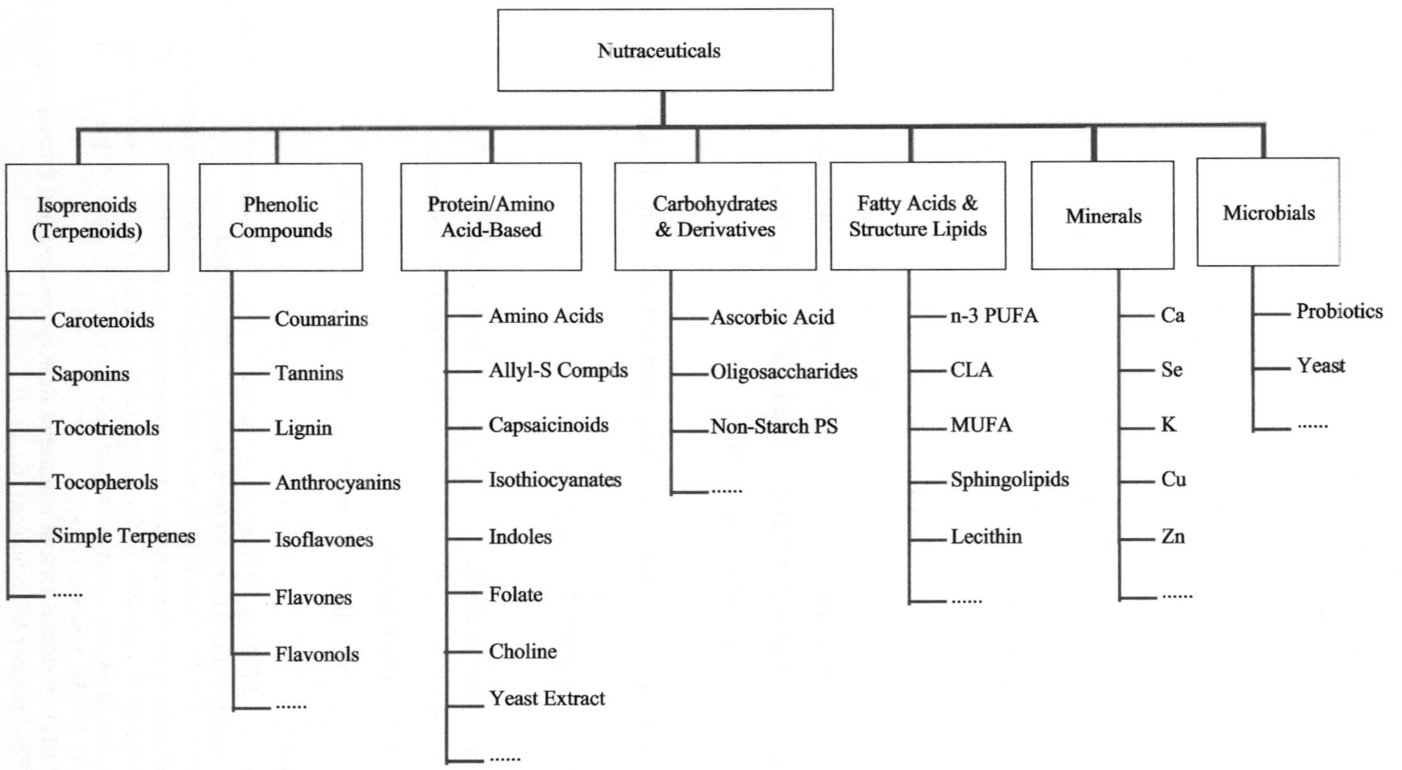

Fig. 2: Schematic Representation of Neutraceutical Compounds.

In conclusion, nutraceuticals are the hybrid of 'nutrition' and 'pharmaceutical'. Nutraceuticals, in broad, are food or part of food playing a significant role in modifying and maintaining normal physiological function that maintains healthy human beings. The principal reasons for the growth of the nutraceutical market worldwide are the current population and the health trends. The food products used as nutraceuticals can be categorized as dietary fibre, prebiotics, probiotics, polyunsaturated fatty acids, antioxidants and other different types of herbal/ natural foods. These nutraceuticals help in combating some of the major health problems of the century such as obesity, cardiovascular diseases, cancer, osteoporosis, arthritis, diabetes, cholesterol etc. Overall, 'nutraceutical' has lead to the new era of medicine and health, in which the food industry has become a research oriented sector.

Key questions

1. Prebiotics were first identified and named by-
   a) Marker b) Rutherford
   c) Marcel Roberfroid d) Stillwell and Lilly
2. Which of the following statement is not correct w.r.t. prebiotics?
   a) Non- digestible food ingredients
   b) They are not drugs
   c) Prebiotics alter the colonic microflora towards a healthier composition
   d) Sensitive to gastric acidity
3. Inulin is a-
   a) Longer chain prebiotic b) Shorter chain prebiotic
   c) Both a and b d) None of the above
4. Explain on Neutraceuticals.
5. Discuss on pharmacological importance of neutraceuticals.

Further reading

Bajury, D. M., Nashri, S. M., King Jie Hung, P., & Sarbini, S. R. (2018). Evaluation of potential prebiotics: a review. Food Reviews International, 34(7), 639-664.

Bindels, L. B., Delzenne, N. M., Cani, P. D., & Walter, J. (2015). Towards a more comprehensive concept for prebiotics. Nature reviews Gastroenterology & hepatology, 12(5), 303-310.

Cardoso, B. B., Amorim, C., Silvério, S. C., & Rodrigues, L. R. (2021). Novel and emerging prebiotics: Advances and opportunities. In Advances in Food and Nutrition Research (Vol. 95, pp. 41-95). Academic Press.

Da Costa, J. P. (2017). A current look at nutraceuticals–Key concepts and future prospects. Trends in Food Science & Technology, 62, 68-78.

Davani-Davari, D., Negahdaripour, M., Karimzadeh, I., Seifan, M., Mohkam, M., Masoumi, S. J., ... & Ghasemi, Y. (2019). Prebiotics: definition, types, sources, mechanisms, and clinical applications. Foods, 8(3), 92.

Gupta, C., & Prakash, D. (2015). Nutraceuticals for geriatrics. Journal of traditional and complementary medicine, 5(1), 5-14.

Gupta, R. C., Lall, R., & Srivastava, A. (Eds.). (2021). Nutraceuticals: efficacy, safety and toxicity. Academic Press.

Lockyer, S., & Stanner, S. (2019). Prebiotics–an added benefit of some fibre types. Nutrition Bulletin, 44(1), 74-91.

Mohanty, D., Misra, S., Mohapatra, S., & Sahu, P. S. (2018). Prebiotics and synbiotics: Recent concepts in nutrition. Food bioscience, 26, 152-160.

Olveira, G., & González-Molero, I. (2016). An update on probiotics, prebiotics and symbiotics in clinical nutrition. Endocrinología y Nutrición (English Edition), 63(9), 482-494.

Souyoul, S. A., Saussy, K. P., & Lupo, M. P. (2018). Nutraceuticals: a review. Dermatology and Therapy, 8, 5-16.

Valcheva, R., & Dieleman, L. A. (2016). Prebiotics: Definition and protective mechanisms. Best Practice & Research Clinical Gastroenterology, 30(1), 27-37.

Valdés, L., Salazar, N., González, S., Arboleya, S., Ríos-Covián, D., Genovés, S., ... & Gueimonde, M. (2017). Selection of potential probiotic bifidobacteria and prebiotics for elderly by using in vitro faecal batch cultures. European Food Research and Technology, 243, 157-165.

Wildman, R. E., Wildman, R., & Wallace, T. C. (2016). Handbook of nutraceuticals and functional foods. CRC press.

Wu, H., & Chiou, J. (2021). Potential benefits of probiotics and prebiotics for coronary heart disease and stroke. Nutrients, 13(8), 2878.

3

Taxonomy of Lactobacilli and Bifidobacteria

Introduction

From earlier chapters, it is evident that Lactobacilli and Bifidobacteria are the two important genera that have potent applications in food microbiology and human nutrition due to their role in food and feed production, preservation and also due to probiotic potentialities. Most of the bacterial species are still unknown indicating that the identification and classification of bacteria is a hard task since the beginning of microbiology and consequently our knowledge about bacterial ecology is very low. Mean while the expectations about novel bacterial products are high. Therefore, bacterial identification has become a growing field of interest in microbiology. Since, Lactobacilli and Bifidobacteria genera include a large number of species and strains exhibiting important properties in an applied context, chiefly in the area of probiotics as food and therapeutics, an updated list of species belonging to these two genera, their phylogenetic relationships and other relevant taxonomic information are of great significance. The classification and identification of a probiotic isolate at strain level point out its typical habitat and origin. Moreover, the species, or genus name, may also indicates the strain's safety and technical applicability for its use in probiotic products. Taxonomy helps to place a newly identified organism in a relevant group and enables the scientific communication. Further it facilitates the construction of databases that can be used for rapid identification of organisms at laboratory level easily. Thus, taxonomy of Lactobacilli and Bifidobacteria helps to study these organisms scientifically and allows characterizing and arranging them in an orderly manner which promote the scientific community to explore the benefits from those organisms for the betterment of mankind.

Taxonomic classification of lactobacilli and bifidobacteria

Taxonomy is a Greek term meaning *taxis*- arrangement or order, and *nomos*- law or to distribute, method.The word taxonomy was first introduced in 1813 by a Swiss botanist Augustin Pyramus de Candolle in his *Théorieélémentaire de la botanique*. Taxonomy is the science of biological classification that describes the groups of organisms on the basis of shared characteristics and giving names to those groups. Bacterial taxonomy can be defined as rank based classification of similar groups of bacteria according to international codes and rules. The Swedish botanist Carolus Linnaeus, regarded as the father of taxonomy for developing a system for naming, ranking and classifying organisms known as Linnaean classification. This renowned system is in common practice for categorization of organisms and binomial nomenclature for naming organisms worldwide. According to Linnaeus classification, each species has to be assigned to a genus, which in turn is a lower level of a hierarchy of ranks which includes family, suborder, order, subclass, class, division/phyla, kingdom and domain. Currently, the classification of life includes three domains, the higher level of hierarchy namely Eukaryotes, Bacteria and Archaea. Most of the probiotic isolates belong to the domain Bacteria.

Classification, identification and nomenclature are the three separate but related sub disciplines of taxonomy. The process of clustering of organisms into taxonomic groups (*taxa*) on the basis of similar characteristics they share or according to a set of criteria is called *classification*. While the process of assigning the names to the taxonomically clustered groups according to internationally accepted rules is called *Nomenclature*. Whereas identification can be defined as a process of determination of the belongings of a new isolate to one of the established and identified taxonomic groups.

The basic unit of biological classification is species which ranks below the genus in hierarchy of taxonomical ranks. It can be defined as a cluster of genomically coherent living organisms, where individuals are capable of exchanging genes or interbreeding and shows a high degree of overall similarity with respect to many independent characteristics. Further, a species can be divided into two or more sub species based on minor but consistent phenotypic and genotypic variations. Therefore subspecies may be considered as the lowest taxonomic ranks in nomenclature for organisms with inter species variations.Phylogenetic aspect and overall similarity of an organism are the two key points to be considered while defining a species.

Methods of taxonomic classification

Conventional or traditional, chemotaxonomic and molecular methods are the three different microbial identification methods practiced in taxonomical studies. Conventionally, bacteria have been classified on the basis of phenotypic and biochemical properties such as growth characteristics, morphology, motility, ability to ferment various sugars with or without gas production, growth at different temperatures, pH, production of organic acids, antibiogram and other biochemical properties. However, comparative 16S ribosomal RNA sequencing analysis based studies showed that some taxa generated on the basis of phenotypic features do not analogous with the suggested phylogenetic relations indicating that some species are not readily distinguishable by phenotypic characteristics. This observation has been proven true for *Lactobacillus acidophilus*, the *Lactobacillus casei* and *Lactobacillus paracasei* groups, and some bifidobacterial strains which have been introduced in many probiotic foods. In last decade, chemotaxonomic and molecular methods have proven efficient and also beneficial in overcoming the limitations of conventional methods. Chemotaxonomy classifies the organisms based on differences and similarities in chemical markers such as analysis of quinine system, fatty acid profiles, polyamine and polar lipid patterns, peptidoglycan diaminoacids, whole cell sugars and cellular protein patterns. Though, chemotaxonomic methods give both qualitative as well as quantitative data, requirement for special apparatus and skills for their successful application confine them to academic purpose and to specialized laboratories. Always progress in bacterial taxonomy has been made with the advances in molecular technology. It can be relevant to consider that the modern bacterial taxonomy has constructed and developed entirely based on the molecular data.

Consequently, modern molecular techniques, including polymerase chain reaction-based and other genotyping methods, have become increasingly important for species identification. Plasmid profiling, DNA-DNA hybridization, DNA-RNA hybridization, G+C content, restriction enzyme analysis (REA), pulse-field gel electrophoresis (PFGE), randomly amplified polymorphic DNA (RAPD), and DNA sequencing of sub-units of 16S and 23S rRNA are commonly used molecular typing methods. These bacterial genotyping methods fall into three categories namely DNA hybridization, DNA sequencing and DNA banding pattern. In case of DNA hybridization dependent methods, hybridization of known probes has been analyzed to discriminate the bacteria. In DNA sequencing methods, bacteria can be distinguished by comparing the sequence of known gene. Whereas in case of DNA banding methods, comparison of restriction digested DNA fragment bands and amplification of known or sometimes unknown region of the genomes helps in the discrimination of bacteria. Earlier, plasmid profiling had been considered best suitable method for the typing of individualsat interspecies level but due to instability of the extra chromosomal DNA, the methods that use chromosomal DNA are preferred over plasmid profiling technique. Restriction enzyme analysis is one such DNA banding method that uses chromosomal DNA to distinguish bacterial isolates. It uses restriction enzymes to cleave the DNA and a gel electrophoresis technique to separate the DNA fragments. But complex banding pattern may pose visual difficulties, however it can be solved using computer aided multivariate analysis. Strains of *L. acidophilus*, *L. casei, L. rhamnosus*, and *L. reuteri* have been successfully differentiated by REA. PFGE is another DNA banding method that can be considered as modification of REA. This method reduces the complex banding pattern using rare cutting restriction enzymes that generate fewer fragments, attributing to its high discriminatory power, enabling intraspecies differentiation. Strains of Bifidobacteria, *L. casei* and *L. acidophilus* were differentiated successfully by PFGE. On contrary, ribotyping is another most widely used DNA banding method which uses ribosomal genes, and is highly discriminative at species level than at the strain level. The banding patterns are more easily interpretable and significant property of this technique is that it employs genes coding for rRNAS i.e. 23S and 16S rRNA which uses universal probes due to the similarities in ribosomal genes of prokaryotes. Gene for 16S rRNA is 1500bp long with 10 highly conserved regions and 10 divergent regions which are highly resistant to mutation. These properties of ribosomal nucleic acids make them phylogenetically important molecules for species identification.

In taxonomic practice, DNA–DNA similarity assays have become the gold standard technique for the delineation of bacterial species. Generally it has been used to measure the genetic distance between two species. But the development of a multi locus sequencing approach has been proposed, potentialities of which have not been widely explored yet. Whole genomic DNA-DNA hybridization (DDH) is not widely used due to difficulties in its practical implementation. However, random genome fragments and DNA microarray technology based methods surmounts the disadvantages of whole genome DNA-DNA hybridization. DNA–DNA hybridization, using 70% relatedness as a threshold for species delineation is applied in most of the earlier phylogenetic studies. DDH values $\leq$ 70% indicate that the organism tested belongs to a different species other than the type of reference organism whereas DDH values $\geq$ 70% show that the test and reference organisms are closely related. Melting temperature (T_m) of hybridized DNA is also useful in determining the genetic relatedness of two different bacterial genomes. During DNA-DNA hybridization process, a sequence of DNA from known organism is considered as reference and sequence of unknown as test. If both the DNA sequences are more similar and/or from more closely related organisms, they form strong bonds during reassociation compared to more dissimilar DNA sequences. Therefore, similar sequences are highly thermostable and require comparatively high melting temperature to denature their hybrid while dissimilar sequences with weak bonds melt at very low temperature. It can be calculated as delta T_m or ΔT_m i.e. change in melting temperature by subtracting the denaturing temperature of heteroduplexDNA from the denaturing temperature of homoduplex DNA. Higher the ΔT_m more distant relationship and lower the ΔT_m more closer relationship between the organisms. However, owing to technical difficulties, DDH may be substituted by the simpler and more accurate technique of comparing average nucleotide or amino acid identities (ANI/AAI) between orthologs of separate genomes. Orthologs are the homologous sequences found in different species that arose from a common ancestral gene during speciation, which may or may not be responsible for a similar function. A species cut-off value of 98% ANI will correspond to 70% DNA–DNA relatedness.

Genomic organization of lactobacilli and bifidobacterium

Whole genomic sequencing for bacterial phylogeny is a laborious process and practically very difficult to perform. To make the classification and identification of new bacterial isolates as a regular laboratory work, the phylogeny of bacteria must be based on the comparison of highly conserved molecules that are present in all microorganisms rather than comparing whole genomes. Therefore, genes encoding rRNA, comprising conserved and variable domains are preferable for phylogenetic studies. Currently, comparision of rRNA sequence is considered to be one of the most powerful and accurate technique for determining the degree of phylogenetic relationship among the microorganisms. Earlier, DNA-rRNA hybridizations or rRNA cataloging methods were used for phylogenetic studies.Technological advances in molecular biology enabled the sequencing of long stretches of rRNA, using reverse transcriptase and direct polymerase chain reaction (PCR) sequencing of 16S or 23S rDNA molecules, resulted in large sequence databases. On the basis of the available information on 16S/23S rRNA sequences, phylogenetic trees or dendrograms were created. Other newly developed molecular typing techniques such as PCR-ribotyping, amplified DNA restriction analysis, rep-PCR, and restriction or amplified fragment length polymorphism, present high sensitivity and discriminatory power for the identification and differentiation of probiotic microorganisms. Though different methods are available for phylogenetic study, each method has its own way to analyze the molecular data; usually of DNA and RNA. Therefore polyphasic taxonomy combining phenotypic, biochemical and physiological characteristics of bacteria with molecular based techniques is useful in accurate identification of a bacterial isolate. Comparison of results obtained from a large number of techniques helps in complete investigation of a microorganism i.e. more the data; better will be the comparisons, such that identification will be more complete and accurate.

Phylogenetic classification of lactobacilli

Lactic acid bacteria (LAB) comprise 13 genera, among which *Lactobacillus* is the largest and extremely diverged genus with more than 180 recognized species. It belongs to the phylum Firmicutes, order Lactobacillales, class Bacilli and family Lactobacillaceae. Lactobacillales is the largest order of the phylum comprising seven families among which Leuconostocaceae is very close to Lactobacillaceae. Lactobacilli are Gram positive anaerobic or microaerophilic rods that constitute major part of lactic acid bacteria, members of which ferment lactose to lactic acid. Due to its vast diversity, taxonomical classification of the genus *Lactobacillus*is not satisfactory. To consider this, lets discuss the classification of *Lactobacillus acidophilus,* one of the most important probiotic species of the genus. Its phenotypic assessment is difficult because of its heterogeneity. In 1960, Lerche and Reuter gave the hint for heterogeneity of the species by suggesting 4 different biotypes. Later in 1980, DNA-DNA hybridization studies assured the heterogeneity, suggesting the existence of 6 distinguished homology groups namely *Lactobacillus acidophilus, L. amylovorus, L. crispatus, L. gallinarum, L. gasseri* and *L. johnsonii*. Consequently, strains of a homology group that showed a high degree of DNA relatedness with *L. acidophilus* remained in this species, while strains of the other homology groups were classified as separate species, i.e. *L. amylovorus, L. gallinarum, L. crispatus, L. gasseri,* and *L. johnsonii*. Although these are regarded as distinct species, they are closely related and have been suggested under one phylogenetic group.

When new genome sequences become available, better phylogenetic inference can be made from a large number of genes or proteins or indeed the whole genome sequence. Genome sequencing not only helps in taxanomical studies, indeed it gives an immense knowledge about a species and variations in their strains, reasons for their existence in specific habitats and metabolic pathways adopted to survive in the particular environment and also for their phenotypic as well as biochemical characteristics. The first *Lactobacillus* genome to be sequenced was that of *L. plantarum WCFS1*. Identification of genes responsible for the fermentation of sugars either by glycolysis or the phosphoketolase pathway rationalized its hetero fermentative nature. Though it has an elaborated peptide degrading proteolytic system, it is capable of *de novo* synthesis of many amino acids. Unlike most of the other species, it is found in wide range of nutrient rich environment. This environmental flexibility attributes to the 600kb region near replication origin which represents a very high number of genes for sugar transport, metabolism and other extra cellular activities. Comparative genomic hybridization of genomic DNA of different strains of *L. plantarum* high light edits genomic diversity and variability.

The second Lactobacillus genome to be fully sequenced was that of the *L. johnsonii* strain NCC533, a human isolate. The strain is a strict anaerobe and is an obligate homo fermentative. In contrast to *L. plantarum,* it lacks the ability of *de novo* synthesis of amino acids. It also lacks genes for the *de novo* synthesis of purine and many other cofactors. Hence, it is mainly found in upper gastro intestinal tract which is rich with amino acids and peptides that are sufficient for its growth. Genomic analysis showed that it has very high number of genes for permeases, peptidases and PTS transporters. In addition, genome analysis of *L. salivarius* subspecies *salivarius* UCC118 revealed a 242 kb megaplasmid, pMP118. The unexpected presence of a megaplasmid is the first verified case among LAB. Genome annotation showed that *L. salivarius* is less auxotrophic than *L. acidophilus* and *L. johnsonii*, since it can synthesize amino acids *de novo* and capable to interconvert nine amino acids. But it is more dependent on extracellular amino acid uptake systems than *L. plantarum.* In contrast to other lactobacilli whose genomes have been sequenced, *L. sakei* is an organism that although a human GI tract transient, is primarily associated with fermented meat and fish. Its evolution to use meat as a growth substrate has given *L. sakei* metabolic capacities that were explained by the genome sequence and that differentiate it from other LAB. For instance, the ability to catabolize arginine, an amino acid abundant in meat, becomes vital for survival when sugar availability is low. This ability is not found among the other sequenced *Lactobacillus* genomes. *L. sakei* also encodes pathways for catabolizing inosine and adenosine. This metabolism is an alternative source for energy production when sugars such as glucose are limiting or absent. The genome analysis further highlighted the remarkable robustness and flexibility to withstand adverse environmental changes by *L. sakei.*

Several studies reported significant phylogenetic incongruence of many of the taxa studied and suggested an unclear phylogenetic path for many *Lactobacillus* lineages. But still it is possible to distinguish the following subgeneric groups:

Group A (*L. acidophilus, L. helveticus, L. bulgaricus, L. johnsonii* and *L. gasseri)*

Group B *(L. salivarius, L. plantarum, L. reuteri, L. brevis* and *P. pentosaceus)*

Group C *(L. sakei* and *L. casei)*

Group D *(Leuconostoc mesenteroides* and *O. oeni)*

Whether theses subgroups possess sufficient phylogenetic entities that serve as basis for reclassification of *Lactobacillus* genus is the huge task. There is relatively little variation between group A and group B; group C is basal to group A and group D is basal, or almost basal, to group B. Moreover, group C and D are populated by only two species each and are not sufficient to form the basis for new genera. In present scenario, there are many challenges to reclassify the genus. Phylogeny is correlated with physiological, environmental and morphological factors. If species are classified only on the basis of genetic relatedness, then that classification may not be satisfactory. Since morphology, physiology and environmental factors influence the phylogenety it's essential to consider all these factors while classifying the genus. In above mentioned sub generic groups, there is no ecologic association among the species. For example, in group A, *L. acidophilus* is found within the gastrointestinal tract whereas its close relative *L. helveticus* is majorly found in cheese. Similarly in Group C, *L. sakei* is an inhabitant of fermenting meat whereas *L. casei* is majorly associated with dairy products. So far proposed phylogenies are mainly based on the ubiquitous genes of the taxa. However, as discussed, phenotypic differences usually arise from differences in gene content. Inclusion of more number of species at key taxonomic position may helps to overcome the phylogenetic difficulties. Based on 16S rRNA phylogenic studies, it can be predicted that *L. reuteri, L. casei, L. sakei, Leuconostoc mesenteroides and O. oeni*, of which the latter four will further populate the dual-inhabitant groups C and D, *L. antri, L. buchneri, L. coleohominis* and *L. fermentum* are likely to be welcome additions to group B and *L. rhamnosus and L. paracasei* to group C.

So far varieties of methods are followed to reclassify the genus *Lactobacillus*. Each study alone can give a hope in reclassifying the genus. But when results of different methods of phylogenic studies of *Lactobacillus* were compared it again leads to ambiguity. Core proteins are found useful in comparative genomic studies. Fig.2 (a) illustrates the phylogenetic relations between different species of *Lactobacillus* based on the sequences of identified 141 concatenated core proteins. Whereas phylogenetic tree in fig. 2(b) was constructed on the basis of four subunits of RNA polymerase sub units and the phylogenetic trees in the fig. 2(c) and fig. 2(d) ware constructed on the basis of the 16s rRNA genes and GroEL proteins repectively. Significant incongruence was found in the trees. However, many studies concluded that GroEL is an important phylogenetic marker for *Lactobacillus* when whole genome information is unavailable. In this chapter, only four subgeneric groups are discussed. Identification and reclassification of a genus is continuous. Reclassifications have been frequent following the identification of new species and classification techniques, and the taxonomy is generally considered unsatisfactory. In recent years advent of new technologies fetched the attention of researchers towards taxonomy of *Lactobacillus*. In future, continuous sequencing and further gene analysis may contribute to the addition of more number of species to the genus *Lactobacillus*.

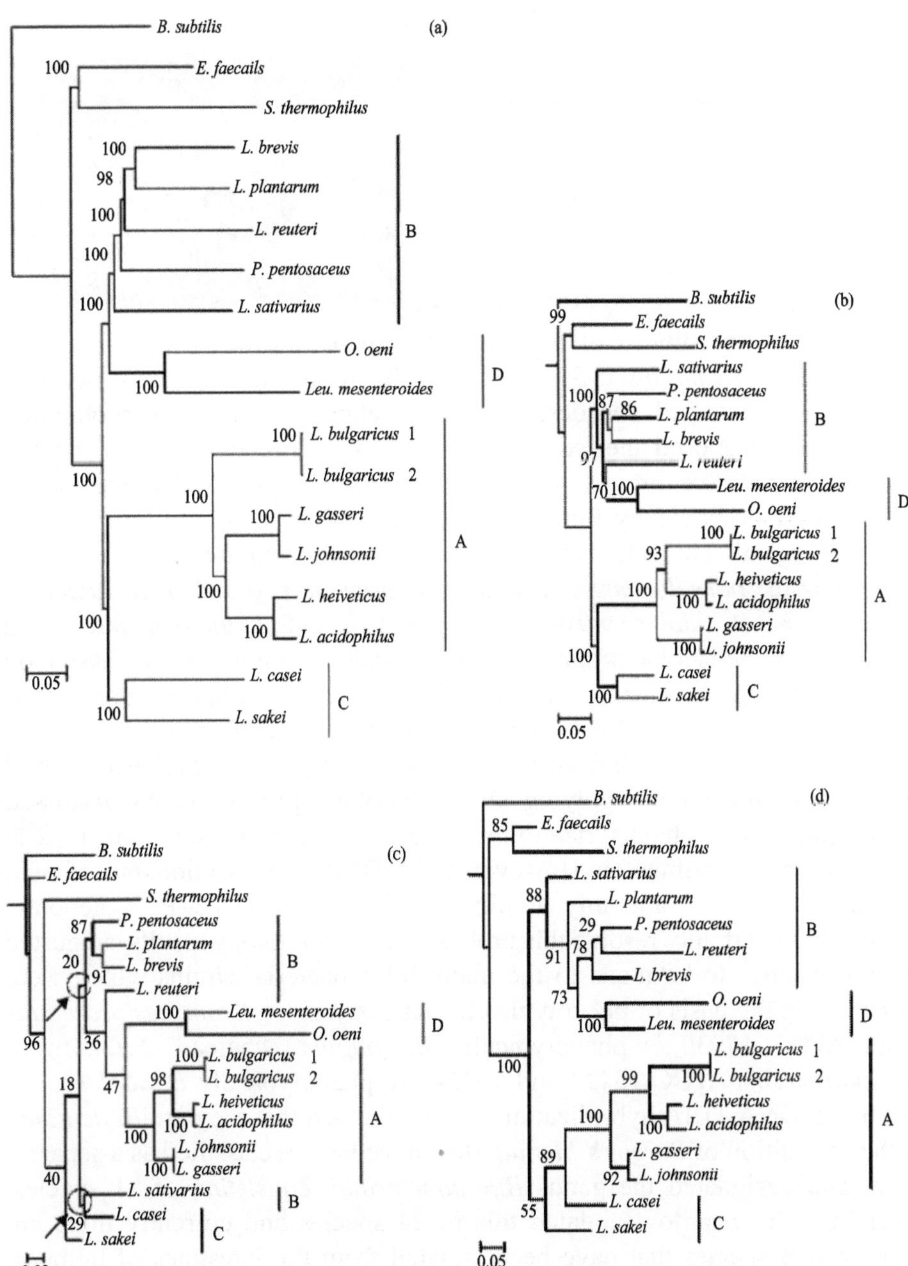

Phylogenetic tree based on:

Fig. (a) sequences of 141 concatenated core proteins, (b) four RNA polymerase subunits, (c) 16S rRNA genes, (d) GroEL proteins

Phylogenetic classification of bifidobacterium

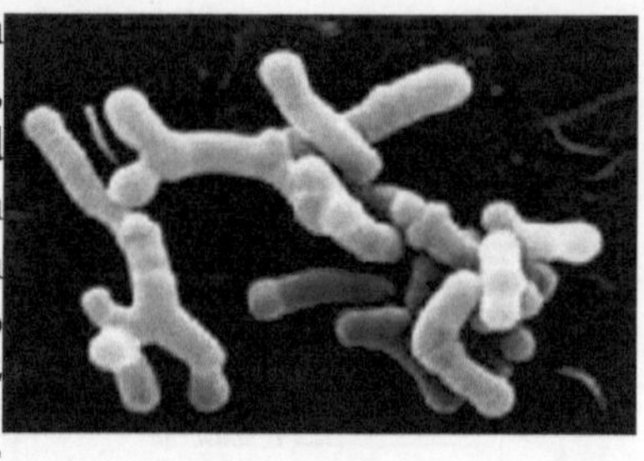

Bifidobacteria were first isolated from the feces of breast-fed infants in 1899, by Henri Tissier, and were designated *Bacillus bifidus*. Even though Orla-Jensen proposed the genus *Bifidobacterium* in 1924, bifidobacteria were classified into other taxonomic groups, such as *Bacillus bifidus*, *Bacteroides bifidus*(1923 to 1934, in the 1st to 4th editions of *Bergey's Manual of Systematic Bacteriology*), and *Lactobacillus bifidus*(1939 to 1957, in the 5th to 7th editions of *Bergey's Manual*), for several decades. In 1957, Dehnert, based on carbohydrate fermentation proposed the existence of multiple biotypes of Bifidobacteria and recognized five groups. In addition to the known *B. bifidum*, seven species were proposed by Reuter on the basis of carbohydrate fermentation and serological characterization: *Bifidobacterium bifidum* var. *a* and*b, B. infantis, B. parvulorum*var. *a* and *b, B. breve* var. *a* and *b, B. lactentis, B. liberorum, B. adolescentis*var.*a, b, c* and *d,* and *B. longum* var. *a* and *b.* Moreover, Mitsuoka proposed a major classification scheme based on physiological and biochemical characterization, which includes four hundred and eighty three strains of Bifidobacteria. In 1971, assessment of the validity of previously described bifidobacterial species at genetic level was started by Scardovi and others applying DNA homology procedure. As discussed earlier, phenotypic characterization may leads to confused state and fails to make accurate identification. However, DNA-DNA hybridization, pulsed field gel electrophoresis, randomly amplified polymorphic DNA–PCR genomic methods will certainly resolve this problem. DNA homology results suggested several changes to be made to the many bifidobacterial strains which were classified on the basis of phenotypic characterization. *Actinomyces arabifidus* strain ATCC 17930 is phenotypically *B. longum,* whereas, *Actinomyces eriksonii* strains ATCC 15423 and 15424 are phenotypically *B. adolescentis* var. b, but DNA-DNA hybridization studies showed that they are *B. dentium.* In the 8th edition of *Bergey's Manual* they have been reclassified as a separate taxon and designated the genus *Bifidobacterium, consisting* of 11 species. Later in 1986, Scardovi updated this to 24 species and currently there are 31 proposed species that have been isolated from the intestines of humans, animals, and insects, and also from human dental caries and raw milk.

Species number	Name	Subspecies	Origin	%G+C
1	*B. adolescentis*		Intestine of adult	59.6 ± 0.8
2	*B. angulatum*		Human feces	59.0 ± 0.1
3	*B. animalis*	*B. animalis*subsp. *Animalis*	Animal feces	60.1 ± 0.3
		*B. animalis*subsp. *lactis*	Yogurt	61.9
4	*B. asteroids*		Intestine of honeybee	59.0
5	*B. bifidum*		Infant feces	62.3
6	*B. bombi*		Intestine of bumblebee	47.2
7	*B. boum*		Rumen of cattle	60 ± 0.2
8	*B. breve*		Intestine of infant	58.8 ± 0.4
9	*B. catenulatum*		Intestine of adult	54.0 ± 0.2
10	*B. choerinum*		Porcine feces	66.3 ± 0.2
11	*B. coryneforme*		Intestine of honeybee	ND
12	*B. crudilactis*		Raw milk	56.4
13	*B. cuniculi*		Feces of rabbit	64.1 ± 0.4
14	*B. dentium*		Human dental caries	61.2 ± 0.4
15	*B. gallicum*		Human feces	ND
16	*B. gallinarum*		Chicken cecum	65.7 ± 1.5
17	*B. indicum*		Intestine of honeybee	60.0
18	*B. longum*	*B. longum*subsp. *infantis*	Intestine of infant	60.5 ± 0.3
		*B. longum*subsp. *longum*	Intestine of adult	60.8 ± 0.8
		*B. longum*subsp. *suis*	Porcine feces	62.0
19	*B. magnum*		Rabbit feces	60.0 ± 0.6
20	*B. merycicum*		Bovine rumen	ND
21	*B. minimum*		Sewage	61.5
22	*B. pseudocatenulatum*		Infant feces	57.5 ± 0.3
23	*B. pseudolongum*	*B. pseudolongum* subsp. *globosum*	Bovine rumen	63.8 ± 0.4
		B. pseudolongum subsp. *pseudolongum*	Porcine feces	59.5 ± 0.4

24	*B. pyschraerophilum*		Porcine feces	59.2
25	*B. pullorum*		Chicken feces	67.5 ± 0.4
26	*B. ruminantium*		Bovine rumen	ND
27	*B. saeculare*		Rabbit feces	ND
28	*B. scardovii*		Human blood	ND
29	*B. subtile*		Sewage	61.5
30	*B. thermophilum*		Porcine feces	60.0
31	*B. thermacidophilum*	*B. thermacidophilum* subsp. *porcinum*	Porcine feces	ND
		B. thermacidophilum subsp. *thermacidophilum*	Sewage	ND

ND-not determined

Currently assigned species of the genus Bifidobacterium (Source: Microbiol. Mol. Biol. Rev. September 2010 vol. 74 no. 3 378-416)

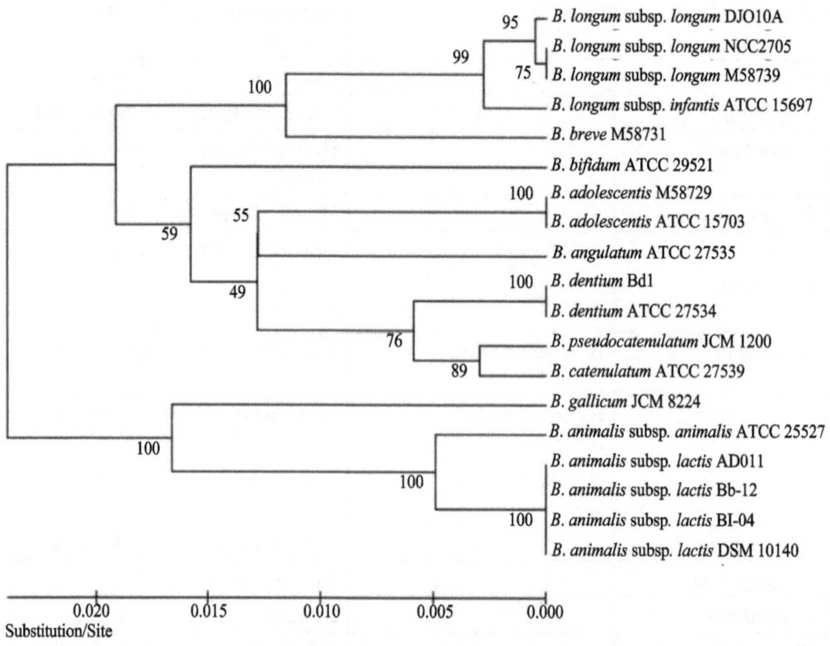

Phylogenetic analysis of the 16S rRNA genes of the nine sequenced bifidobacteria, illustrating their separation into three distinct groups. Completely sequenced *Bifidobacteria* are indicated in bold. (source: Lee J , and O'Sullivan D J. Microbiol. Mol. Biol. Rev. 2010;74:378-416).

Diversity of an organism is a root cause for its taxonomic studies. Taxonomy not only assigns names to the identified organism. Since classification and identification are essence of taxonomy, characterization is required. Characterization helps to understand the in and out of an organism thoroughly which in turn helps to exploit beneficial aspect of that isolate for the betterment of mankind. As discussed above, the genomic studies explain the basis for diversity of *Lactobacillus* and *Bifidobacterium* species. Understanding diversity fulfills the needs of taxonomy.

Today taxonomy is often based on a polyphasic approach which involves genotypic and phenotypic methods. Pure historically phenotypic methods have dominated the identification and classification scheme of Lactobacilli. In recent years, 16S rRNA sequencing has become the method of choice, not only because of its high degree of portability, but equally important because of the availability of a large database of reference sequences. In a polyphasic approach, both phenotypic and genotypic methods are involved. It may leads to the differences in the phylogenetic trees based on the 16S RNA sequencing and the phenotypic groups based on fermentation profiles and metabolic activity. However, 16S rRNA sequencing has some considerable disadvantages, a polyphasic approach that involves a balanced use of multiple genotypic and phenotypic methods will always yield a more balanced and reliable result. As sequencing become more accessible and cheaper, the role of a single molecule such as 16S rRNA tend to fade. The sequencing of variety of individual household or housekeeping genes will diversify and broadened the taxonomic views with the sequencing of the complete genome of many organisms as a feasible option in near future.

Key questions

1. The word taxonomy was first introduced by

   a) Augustin pyramus de candolle b) Carolus Linnaeus

   c) Walter Rosen d) Historia Animalium

2. Basic unit of biological classification is.............Species

   a) Genus b) Species

   c) Variants d) Order

3. Lerche and Reuter in 1980 contributed the

   a) Identification of genes responsible for fermentation.

   b) Classification scheme for bifidobacteria based on physiological and biochemical characterization.

   c) Heterogeneity of the species by suggesting 4 different biotypes.

   d) Chemotaxonomic classification.

4. The genus bifidobacterium was proposed by
   a) Orla-Jensen b) Dehnert
   c) Mitsouka d) Scardovi
5. Molecular methods of lactobacillus and bifidobacteria classification.
6. Phylogenetic tree construction of bifidobacterium.

Further reading

Bezkorovainy, A. (2020). Classification of bifidobacteria. In Biochemistry and physiology of bifidobacteria (pp. 1-28). CRC Press.

Feyereisen, M., Mahony, J., Kelleher, P., Roberts, R. J., O'Sullivan, T., Geertman, J. M. A., & van Sinderen, D. (2019). Comparative genome analysis of the Lactobacillus brevis species. BMC genomics, 20, 1-15.

Goldstein, E. J., Tyrrell, K. L., & Citron, D. M. (2015). Lactobacillus species: taxonomic complexity and controversial susceptibilities. Clinical Infectious Diseases, 60(suppl_2), S98-S107.

Hill, D., Sugrue, I., Tobin, C., Hill, C., Stanton, C., & Ross, R. P. (2018). The Lactobacillus casei group: history and health related applications. Frontiers in microbiology, 9, 2107.

Huang, C. H., Li, S. W., Huang, L., & Watanabe, K. (2018). Identification and classification for the Lactobacillus casei group. Frontiers in Microbiology, 9, 1974.

Lu, W., Pei, Z., Zang, M., Lee, Y. K., Zhao, J., Chen, W., ... & Zhang, H. (2021). Comparative genomic analysis of bifidobacterium bifidum strains isolated from different niches. Genes, 12(10), 1504.

Lugli, G. A., Milani, C., Duranti, S., Mancabelli, L., Mangifesta, M., Turroni, F., ... & Ventura, M. (2018). Tracking the taxonomy of the genus Bifidobacterium based on a phylogenomic approach. Applied and environmental microbiology, 84(4), e02249-17.

Mattarelli, P., Biavati, B., Holzapfel, W. H., & Wood, B. J. (Eds.). (2017). The Bifidobacteria and related organisms: biology, taxonomy, applications. Academic Press.

Mattarelli, P., Holzapfel, W., Franz, C. M., Endo, A., Felis, G. E., Hammes, W., ... & Dellaglio, F. (2014). Recommended minimal standards for description of new taxa of the genera Bifidobacterium, Lactobacillus and related genera. International Journal of Systematic and Evolutionary Microbiology, 64(Pt_4), 1434-1451.

Michie, S., Wood, C. E., Johnston, M., Abraham, C., Francis, J., & Hardeman, W. (2015). Behaviour change techniques: the development and evaluation of a taxonomic method for reporting and describing behaviour change interventions (a suite of five studies involving consensus methods, randomised controlled trials and analysis of qualitative data). Health technology assessment, 19(99), 1-188.

Murali, A., Bhargava, A., & Wright, E. S. (2018). IDTAXA: a novel approach for accurate taxonomic classification of microbiome sequences. Microbiome, 6, 1-14.

O'Callaghan, A., & Van Sinderen, D. (2016). Bifidobacteria and their role as members of the human gut microbiota. Frontiers in microbiology, 7, 925.

Sanders, J. W., Oomes, S. J. C. M., Membré, J. M., Wegkamp, A., & Wels, M. (2015). Biodiversity of spoilage lactobacilli: phenotypic characterisation. Food microbiology, 45, 34-44.

Simon, H. Y., Siddle, K. J., Park, D. J., & Sabeti, P. C. (2019). Benchmarking metagenomics tools for taxonomic classification. Cell, 178(4), 779-794.

Usman, M., Britto, R., Börstler, J., & Mendes, E. (2017). Taxonomies in software engineering: A systematic mapping study and a revised taxonomy development method. Information and Software Technology, 85, 43-59.

Wittouck, S., Wuyts, S., Meehan, C. J., van Noort, V., & Lebeer, S. (2019). A genome-based species taxonomy of the Lactobacillus genus complex. MSystems, 4(5), 10-1128.

Yang, B., Chen, Y., Stanton, C., Ross, R. P., Lee, Y. K., Zhao, J., ... & Chen, W. (2019). Bifidobacterium and Lactobacillus composition at species level and gut microbiota diversity in infants before 6 weeks. International journal of molecular sciences, 20(13), 3306.

4

The Ecology of Lactobacilli in the Gastrointestinal Tract

Introduction

Preceding chapters discuss the essentials of probiotics, their role in maintaining gut health in combination with prebiotics and significance of systematic classification of probiotic isolates. Further, to draw an idea how probiotics function in the gut, it is essential to know the interaction of probiotics with other microorganisms in the gastrointestinal milieu - the site of their probiotic action. Knowledge in the ecology of probiotic microorganisms in host environment such as gut gives an ideal picture of their functioning manner and the factors (both host and microbial) responsible for it.

Study of microbial ecology in the gastrointestinal tract is very difficult since gut harbors highly diversified microflora that varies not only at species level but also at individual level. The diversity of gut microflora in two healthy individuals of same species may exhibit variations in their composition and it may be influenced by geodemographic factors also. In this regard, the present chapter deals with the study of ecology of *Lactobacillus*- most commonly used probiotic bacteria in the gastrointestinal tract which in turn helps in framing selection criteria for screening of potential probiotics.

Recent studies on interaction between human and gut microflora discussed that co-evolution between microorganisms and their host have contributed for the development of a very complex microbial diversity in the gut. Highly diversified microflora in the human gut is a consequence of natural selection at two levels, top-down and bottom-up selection pressure. Hierarchy theory expounds that higher levels compel lower levels for possible organizational solutions i. e. organisms at the higher levels govern the structure or population dynamics of an ecosystem while on contrary lower levels control the structure of ecology when bottom-up level is the selection pressure. In case of interaction between human and gut microflora, being host human is at higher level in the hierarchy for selection and as mentioned above, top down selection pressure is driven by higher levels; host selects the beneficial microorganisms for functional redundancy leading to the development of a microbial community with divergent microbial divisions. Chances for lateral gene transfer also thought to influence the selection process and there by structure of the microbial community. On the other hand, competition among the members of host selected microbiota exert bottom up selection pressure resulting in the development of highly specialized genomes with functionally prominent set of genes. Natural selection at host level obviously favors functionally constant microbial communities which benefits the host with functional redundancy encoded in their genomes. Mean while bottom-up cell-level selection pressure aspire to favor niche specialization to avoid competition between the members of microbial community. In order to be a part of specialized niche, to become more potent microorganisms adapt a particular set of abiotic and biotic factors within the habitat by acquiring new sets of genes essential for their survival and thus escape the competition from decrepit microbial species. Horizontal gene transfer often referred as lateral gene transfer is an important factor in the evolution of microorganisms and is considered as a fundamental process for the evolution of microbial genome. Mechanism of lateral or horizontal gene transfer in prokaryotes confers new characteristics to microorganisms, allowing them to adapt factors essentials for their survival in the new environment and also enhance their metabolic flexibility enabling the utilization of wide range of new compounds in the habitat. If the microorganism has to exist in a new environment, dominating other members of the community, then it has to adapt and utilize the available compounds as its metabolic fuel for oxidation to generate energy required for the maintenance of its life cycle.

Interaction of lactobacilli with gut epithelial cells

In human body, gut is considered one of the major ecological niches where the resident gut flora catalyzes the biotransformation of ingested food. Among the beneficial bacteria of gut, species belonging to the genera *Lactobacillus* and *Bifidobacteria* predominates and exert a beneficial effect on the host by their mechanism of probiotic action. Energy cycling and its recovery from the environment play a significant role in growth of all living organism. Glycolysis being the most important carbohydrate metabolic cycle constitutes the primary energy generating pathway of most of the species of lactic acid bacteria including lactobacilli. Studies on *Lactobacillus* species isolated from mammals showed that glycolytic genes are encoded by central glycolytic operon namely: the gap operon. Gap operon encodes for a putative central glycolytic gene regulator (cggR) and four enzymes- glyceraldehyde-3-phosphate dehydrogenase (gap), phosphoglycerate kinase (pgk), triose phosphate isomerase (tpi) and an enolase (eno). This operon arrangement was first noted in the genomes of *Lactobacillus plantarum* NC8 and *Lactobacillus sakei* Lb790. Later the complete arrangement of this operon has been identified in other species of *lactobacillus* also but in some species the arrangement was found partial. As discussed above, this observation suggests that, top-level selection pressure for functional redundancy and bottom-up level for niche specialization influenced the microorganism and therefore, through evolution and adaptation, mammalian lactobacilli optimized the gap operon gene arrangement for superfluous functionality.

Mechanism of metabolism in lactobacilli

Since diet of man contains a variety of sugars some of which are not even digested by gastric enzymes, lactobacilli in the gut utilize both Embeden-Meyerhof pathway (EMP) and Phosphoketolase Pathway (PKP), which facilitates the utilization of wide range of sugars in the niche. The fermentation pattern for lactobacilli species reveals that some are facultative heterofermenters while few are obligate. Both obligate and facultative heterofermentative lactobacilli express genes for EMP and PKP glycolytic pathways. In general, homofermentative lactobacilli through Embeden-Meyerhof pathway converts carbohydrates into lactate while heterofermentative lactobacilli produce lactate, CO_2 and ethanol by Phosphoketolase Pathway. However, metabolic flux analysis revealed that PKP serve as key glycolytic pathway while EMP acts as a shunt. The use of both the pathways seems to be an adaptation in order to balance the energy yield. Among these two glycolytic pathways, PKP yields less ATP than the EMP and is usually used to ferment pentoses. It is known that homofermentative lactobacilli prefer EMP to meet its energy requirement by fermenting carbohydrates to lactate. But when there is a carbohydrate rich environment, it act as facultative heterofermentative and switches to PKP which yields less ATP. For example, *L. reuteri*, a facultative hetero-fermentative able to utilize both the EMP and PKP to ferment carbohydrates. It sugest that *L. reuteri* and other heterofermentative lactobacilli switches to PKP as an adaptation.

When Prebiotics are in excess use, they act as modulators of intestinal health. This constrains the gut microbiota to be under constant pressure to adapt them to the variety of food that we ingest regularly. In this context, lactobacilli present in the GT of human have developed an array of adaptations at genome level that help to their continued presence as intestinal microbiota.

Often the metabolic flexibility of an organism is determined on the basis of its genome size. Studies correlating the genome size and metabolic flexibility argued that bacterial species with considerably larger genomes are usually capable of fermenting a wider range of carbohydrates which serve as prebiotics while bacteria with relatively smaller genomes found in association with restricted niche habitats. For example, *L. plantarum* has one of the largest genomes among the species of lactobacilli and other lactic acid bacteria sequenced to date. This species of *Lactobacillus* has been isolated from wide range of sources like soil, vegetables, dairy, silage, meat, GT of human as well as other mammalian animals. It is facultative heterofermentative *Lactobacillus* that uses PKP as major glycolytic pathway for the extensive utilization of a wide range of carbohydrates. Its large genome has genes for encoding PTS transporters, ABC (ATP Binding Casette) transporters and glycosyl hydrolases and thus contributing to the metabolic flexibility. In addition to lactate, it produces acetate under anaerobic condition and capable of converting malate, tartrate and citrate to lactate or acetate. It also able to deaminate arginine to ornithine and serine to pyruvate further, pyruvate can be converted to both L and D lactate. Moreover, it has a unique characteristic to tolerate low pH that enables it to survive in the fermenting environment with acidic pH. Therefore, it is one of the dominant species found in a wide range of fermented foods including sour dough, sauerkraut, olives, kimchi, ogi, meat. In contrast, *L. gasseri* with small genome is unable to ferment sugar alcohols, pentoses or deoxysugars, and grow in the habitats devoid of oxygen only. The inability to ferment pentose sugars is due to the absence genes for the two key enzymes transaldolase and transketolase of the pentose phosphate pathway.

Role of nutritional metabolism in genome organization

When an organism is allowed to grow in a nutrient rich environment, sometimes it leads to genome decay due to the loss of genes. In nutrient rich environments where only few carbon and other sources are present in abundant, microorganisms use only few metabolic process to utilize those sources. Upon continuous passage in such environment, it leads to genome contraction and loss of unused carbohydrate transporters and enzymes gradually. For example, the majority of the pseudogenes in *L. helveticus* commonly found in milk are associated with carbon and amino acid metabolism and transportation. It suggest that, since the organism grow in nutrient rich medium like milk, the metabolic process, involving the genes considered as pseudo are less required.

Reproduction in lactobacilli

As mentioned in the beginning of this chapter, Horizontal gene transfer (HGT) has been considered as a significant mechanism responsible for exchange of bacterial genes that leads to the acquisition of new characters. Transposons, bacteriophages and plasmids are well known mobile genetic elements that actively take part in HGT. Plasmids are very common in *Lactobacillus* species also. Studies on plasmids of Lactic acid bacteria estimate that about 38% of *Lactobacillus* species contain one or more plasmids of different size. These plasmids harbor genes mainly for carbohydrate, citrate and amino acid utilization, production of various peptides like bacteriocins and other defensives, stress tolerance and antibiotic resistance. Genes responsible for biosynthesis contribute to the metabolic adaptation through HGT, particularly plasmid carried carbohydrate genes are important mobile genetic elements for lactobacilli. For example, in a study on *L. rhamnosus* Lc705 isolated from gastrointestinal tract, a 64kb plasmid that encodes proteins for the fructose PTS, aldohexoseuptake proteins, a glycosyl hydrolase and a number of other genes for alpha and beta lactoside utilisation and transport was identified. Based on these observations, it is assumed that the presence of plasmid borne genes enabled *L. rhamnosus* Lc705 to compete with other gastro intestinal tract lactobacilli species harboring these genes in their chromosome.

In the last two decades, majority of studies on gut lactobacilli have been focused on the lactobacilli behaviorism in the gastro intestinal tract which are introduced as probiotics and their *in vitro* characterization to ensure safety and efficacy of the isolates. However, research on the ecology of lactobacilli which are naturally present within the gastrointestinal tract of humans and animals, constituting a complex gut microbiota has received much less attention, but it is a crucial aspect of probiotic research and development. Advances in molecular biology technologies have made the detection and identification of lactobacilli easier and allowed researchers for insight into the occurrence, diversity and dynamics of intestinal *Lactobacillus* population. In combination with biochemical and genetic approaches, animal models have been used to disclose the influence of lactobacilli on the host, on other members of gut microbiota and their effect on host, and also to identify some bacterial factors that are responsible for the ecological performance of lactobacilli in the intestine. Functional and comparative genomics have already been carried out for some *Lactobacillus* strains in order to identify features essential for gut colonization and host-microbe co-existence. Functional genomics examines the way in which the genome functions, while comparative genomics helps to compare and identify the differences and similarities in the genomes of the different microorganisms. Research in microbial genomics to decipher these ecological interactions is rewarding because it not only enhance fundamental knowledge about the gut ecosystem and physiology of bacteria in relation to host welfare but also provides information to select lactobacilli for probiotics and other pharmaceutical applications.

Mechanism of adhesion of lactobacilli to the gastrointestinal epithelial cells.

Animal and microbial interactions are highly complex which mainly includes pathogenic, symbiotic and competitive interactions. All these interactions enormously vary and can be found in animal body, specifically the gastrointestinal tract. The animal gastrointestinal tract is divided into various well-defined anatomical regions, it begins at the mouth and includes the esophagus, stomach (rumen-reticulum, crop, gizzard), small intestine (duodenum, jejunum, ileum), and large intestine (cecum and colon), rectum and anus. The main function of the gastrointestinal tract is the conversion of food into simple components which can be easily digested and efficiently absorbed by the host, eliminating waste products.

On the other hand, the gastrointestinal tract is also a complex ecosystem in which host cells and microorganisms interact and is considered as most metabolically adaptable and swiftly renewable organ of the body, which influences the normal nutritional, physiological and immunological functions of the host. However, not all the cells of gastrointestinal tract directly take part in microbial interaction, only single thin layer of epithelial cells are encountered by microorganisms during the interaction. Microbial density in the gut is highly diversified and it differs from one individual to other and collectively they outnumber the total number of animal cells by a factor of 10 i.e. microorganisms in gut are tenfold higher in number compared to total number of cells that constitutes the host. Gastrointestinal tract harbors mainly bacteria and fungi, however, bacteria are the most widely studied group of microorganisms in the gastrointestinal tract. In gut, they are provided with undigested dietary components which escape from gastric digestion in the stomach resisting hydrolytic enzymes and gastric stress, and also the host-derived compounds such as mucopolysaccharides, mucins, epithelial cells, and enzymes. Knowing about the organism living within the other, utilizing its sources and in turn affecting the host organism beneficially and sometimes adversely is really curious.

Several efforts have been made to isolate bacteria from gastrointestinal tract in order to study their function. Earlier, isolation and *in vitro* growth of anaerobic bacteria from the gastrointestinal tract was challenging. However, introduction of the role tube method by Hungate and development of anaerobic glove box initiated a rapid research involving the isolation and characterization of many novel anaerobic bacteria particularly from human gastrointestinal tract. Since then exploration of gut flora continued and bring tremendous changes in the field of microbial genomics. However, with these conventional cultivation methods only a minor portion of the gastrointestinal tract microorganisms, which are cultivable were isolated in the form of pure culture and there were no techniques to isolate and study non cultivable microorganisms. In reality a very small portion of gut microbiota are cultivable; for rest instead of considering as non cultivable, it is relevant to say that the methods for their isolation and cultivation are yet to develop or under development. Recently, this lacuna has been surmounted to some extent with the development of new research area called “Molecular Microbial Ecology” (MME). MME studies majorly focused on the isolation and characterization of non cultivable bacteria. Advances in the molecular biology techniques contributed a lot to the microbial genomics and proteomics, which opened new areas of research in microbiology. Development of anaerobic culture techniques, use of rodent experimental models and the development of gnotobiotic technology (by which germ-free or animal models with a defined microbial community could be derived and maintained) have been served as the source for technological impetus for major advances in understanding the gastrointestinal ecology during the last few decades.

Several factors such as pH, peristalsis, redox potential, bacterial adhesion, bacteria-bacteria interactions, mucus and bile secretion, immune globulins, intestinal enzymes, exfoliated epithelial cells, nutrient availability, diet, as well as bacterial antagonism decides the prevalence of bacteria in different parts of the gastrointestinal tract. In general, number of bacteria in the gastrointestinal tract majorly depends on intestinal pH, bile and peristalsis. In the stomach and the upper part of the small intestine which includes duodenum and jejunum, only 10^3- 10^7bacteria per g of contents are present. It may be due to the acidic environment of the stomach and also the fast peristaltic movement in upper parts of the alimentary canal. On the contrary, in the distal small intestine, ileum and large intestine, a rise in the pH with decreased peristaltic movement favors higher number of bacteria. In gastrointestinal tract lactobacilli and bifidobacteria are predominant beneficial bacterial species.The acid tolerant lactobacilli are one of the dominating species in the stomach, duodenum and jejunum. Lactobacilli feed almost exclusively on sugars, variety of micro nutrients and prefer microaerophilic conditions to grow. However, digestion and absorption of sugars and other complex food components in small bowel provide nutrient rich environment for lactobacilli and less oxygenic condition particularly in the terminal ileum favors the luxuriant growth, making the small bowel an ideal habitat for lactobacilli.

Compared to small bowel, the undigested food entering the large bowel with a narrow range of nutrients confines the lactobacilli number. It does not mean that large bowel is devoid of lactobacilli; it harbors but they are outnumbered by other bacterial genera. In spite of the distribution of microorganisms in different organs, one of the organs that harbors beneficial bacteria come to our mind is colon. It is well known that colon is the largest bacterial reservoir in the body and contains up to 10^{12} bacteria per gram of contents. As mentioned earlier, the primary site of interaction between host and gut microbiota is the intestinal epithelium, which is one cell thick and forms a physical barrier between gut contents and the immune cells. The epithelium consists of distinct epithelial cells, the most predominant among which is the enterocyte. Digestion and barrier action are the two important functions of the enterocyte. Secreted mucins coat the epithelial cells of stomach, intestine and colon and thereby restrict bacterial access to epithelial cells. Therefore, adherence of bacteria to the epithelium is critical for probiotic function.

Adherence to intestinal epithelia and/ or mucus, particularly in the upper digestive tract aids lactobacilli to resist the peristaltic movement and allows intestinal colonization. Studies on probiotic characterization of bacteria revealed that the surface proteins and to lesser extent carbohydrates have a key role in adhesion to the intestinal cells and also in cell-cell communication between bacterial population. It has been also reported that some probiotic bacteria modulate gene expression of the host that could enhance the barrier function of the mucus layer and is beneficial to the host. Because, enhanced function of mucus layer reduces the chances of binding of harmful bacteria to the enterocytes, facilitating their elimination by peristaltic movement. However, even in the presence of enhanced mucus layer function probiotic bacteria with specific surface proteins and carbohydrates can bind easily to the epithelial layer. Once probiotic bacteria start colonizing the enterocytes in epithelial layer, they start occupying available niches, producing regulation factors such as organic acids, hydrogen peroxide and bacteriocins; thereby promote gut maturation and modulate non specific immunity of the host to maintain intestinal immune homeostasis.

Colonization of lactobacilli in gut

Anatomical differences in the digestive tract of mammals greatly influence the composition of lactobacilli. The non-secreting stratified squamous epithelia in the fore-stomach of pigs, mice, rats and horses are efficiently colonized by lactobacilli. In the fowl intestinal tract, the dominating *Lactobacillus* species are *L. crispatus, L. gallinarum, L. johnsonii, L. salivarius* and *L. reuteri.* In fowls, lactobacilli efficiently colonize the stratified squamous epithelium that coats the crop, a pouch in the middle of oesophagus used for the storage of food prior to digestion. Due to their fermentation activity, the colonizing lactobacilli produce acid and lower the pH of crop to 4.5-6.

As discussed above adhesiveness is important in probiotic action. When lactobacilli were treated with proteases resulted in the loss of adhesiveness leading to the assumption that surface proteins mediate the adhesion of lactobacilli in the host gastrointestinal tract. However, several studies also reported that certain carbohydrates and lipoteichoic acids also take part in adhesion. Some of the lactobacilli adhesive factors are characterized at molecular level including the collagen binding CnBP of *L. reuteri,* the collagen and laminin-binding CbsA of *L. crispatus,* fibronectin binding SlpA of *L. brevis* and the pig and hen mucus-binding Mub of *L. reuteri.* ABC transporter operon in *L. reuteri* encodes 29 kDa surface protein CnBP. Whereas CbsA is a 43-kDa S-layer protein consist of single subunit that forms a regular crystalline structure around the cell, constituting the major cellular protein in *L. crispatus.* Molecular characterization of CbsA showed that N-terminal domain possess specific adhesive sequences for collagen type I and IV as well as laminin while C-terminal domain binds to bacterial cell wall. Like CbsA, SlpA is also a S-layer protein that binds *via* its N-terminal domain to immobilized fibronectin and cultivated human intestinal cell lines possible through fibronectin bridging mechanism. Further, a 29-kDa surface protein of *L. fermentum* that binds to small intestinal mucus and gastric mucins in pigs has also been identified. In addition, preliminary studies observed the expression of 15-kDa fibronectin binding protein and a 45-kDa and a 58-kDa protein that bind to collagen type I in *L. acidophilus.*

Antagonistic, Antimicrobial and Immunological properties of lactobacilli

Antagonistic property is another most important beneficial aspect of probiotics that protects the host from pathogenic bacteria. *In vitro* and *in vivo* studies for evaluation of the probiotic potentiality and characterization of lactobacilli using host cell lines and animal models proventhat lactobacilli affect the prevalence of several intestinal pathogens and thereby reduces the risk of infection in the host. For instance, adhesion of *L. acidophilus* to the chicken ileal epithelium excluded *Salmonella pullorum in vitro.* Lactobacilli efficiently excluded *E.coli* and *Salmonella* in human and piglet mucus.

It has also been reported that the expression of host genes responsible for innate immunity can be altered by some of the lactobacilli. It is thought that probiotics in the gut lumen are recognized and processed by the specialized cells of the immune system through several routes.

- Specialized epithelial cells called M (microfold) cells present in the villi or in the follicle-associated epithelium that coats Peyer's patches directly take up probiotics by transcytosis. Engulfing probiotics, the macrophages or dendritic cells (DCs) which are present below the micro fold cells also trigger the host immune responses.

- Through the gaps between the intestinal epithelial cells (IECs), the dendrites of the DCs in the intestinal lamina propria have been found to sample and process probiotics in the gut lumen, extending their dendrites.
- Once they are sampled by immune cells, probiotics directly influence IECs to secrete an array of cytokines, which in turn modulate the immune functions of DCs, T cells, and B cells in the gut-associated lymphoid tissue (GALT).

A variety of cytokines are secreted by probiotic sampled macrophages and DCs. Most significant cytokines among these are IL-12 and IL-10, which are important in regulating the immune response because IL-12 augments cellular immunity whereas IL-10 accounts for the suppression of inflammatory responses. Comparative analyses have revealed that induction for the macrophages to produce IL-12 by lactobacilli strains varies greatly. Upon exposure to *L. salivarius,* a macrophage inflammatory protein 3A (MIP3A) also known as CCL20 was upregulated 38.2 and 20.0 folds as measured by microarray and quantitative PCR respectively. CCL20 is a small cytokine of CC chemokine family that possesses antimicrobial properties since it is chemotactic for lymphocytes and neutrophils. Induction of CCL20 in the epithelial cells due to the interaction with *L. salivarius* gives a clue that how interaction with lactobacilli enhances antimicrobial properties of the host. *L. salivarius* also known to up regulate the expression of genes for chemokines CXCL1 (13 and 47-folds by microarray and quantitative PCR respectively) and CXCL2 (6.4- and 4.5-fold in the microarray array and quantitative PCR data, respectively).

Both of these chemokines proven to have inherent antimicrobial properties similar to those of CCL20 and also actively take part in the prevention of infection in the lung epithelium. In addition, following an exposure to *L. salivarius* also upregulated the expression ofTNFAIP3 gene that encodes for an immune regulator responsible for negative regulation of NF-κB (a protein complex that controls the transcription of DNA) and apoptosis.Moreover, other two regulatory genes of NF-κB namely NFKBIA and BIRC3 were up regulated by 5.4 and 6.1 folds respectively upon an interaction with *L. salivarius*. NFKBIA is one of the most important negative regulators of NF-κB that interact directly with the subunits of NF-κB transcription factor p65 p50 heterodimers, sequestering them in the cytoplasm and preventing their translocation to the nucleus to aid DNA transcription.

Whereas BIRC3 (cIAP2) gene, in contrast, is an NF-κB target gene that is an inhibitor of apoptosis and expression of which in epithelial cells is likely to protect against cytokine-induced apoptosis. Meanwhile *L. casei* YIT 9029 found to augment innate immunity and down regulate inflammatory responses by altering the cytokine profile of the host. In addition, *L. planatarum* another probiotic species known for its extensive metabolic flexibility also induces IL 10, an anti inflammatory cytokine. Studies showed that teichoic acid in its cell wall is responsible for the effective induction of IL 10 in the host intestinal epithelial cells. Whereas in atopic dermatitis, *L. delbrueckii* ssp. *bulgaricus* proven to suppress inflammation by decreasing the levels of IL-6. However, *L.casei*, shown to inhibit antigen-induced IgE production through induction of IL-12 secretion by macrophages, which could be useful in preventing IgE mediated hypersensitivity. Species of lactobacilli with proven anti inflammatory properties either in the form of capsules or along with food adjuncts would be promising in alleviating the symptoms of various inflammatory diseases.

From earlier chapters we know that immune modulation is one of the important probiotic properties. Probiotic in the gut interacts with the various immune cells and thereby influences the host immunity. Out of many immune modulatory functions of probiotics, balancing Th1 and Th2 immune response is of more significant. Th1 and Th2 are the two types of effector $CD4^+$ T helper cells and are designed to eliminate distinct types of pathogens.Th1 accounts for the production of pro-inflammatory cytokines such as interferon gamma (IFN-γ), tumor necrosis factor α (TNF α) and interleukin 2 (IL-2) that activates the bactericidal activities of the macrophages and further induces B cells for opsonization and complement fixation, leading to the cell mediated immunity in the host. Whereas, Th2 immune response results in the release of IL-4, IL-5, IL-6 and IL-13 which activates B cells to produce neutralizing non cytolytic antibodies, leading to humoral mediated immunity. Cytokines produced by Th1 cells stimulate the phagocytosis and destruction of microbial pathogens. Therefore, Th1 immune response is meant for intracellular pathogens while Th2 for extracellular pathogens.

Dendritic cells (DC) a distinct group of antigen presenting cells that are considered as gatekeepers of immune response and principal stimulators of naive T helper cells. DC are known to play a pivotal immune regulatory role in the Th1, Th2, and Th3 cell balance and are present throughout the gastrointestinal tract. Therefore, dendritic cells may be target for modulation by gut microbes, including ingested probiotics.It has been widely suggested that the delicate balance between Th1 and Th2 immunity is pivotally controlled by the stimulating DC. *Lactobacillus* spp. differentially induces IL-6, IL-10, IL-12, and TNF-α. Moreover, lactobacilli induce cytokine production in species and dose dependent fashion. *L. reuteri* is a poor inducer of IL-12, TNF- α and IL-6 when compared with *L. casei.* However, *L. casei* strongly upregulates MHC class II than any other species of *Lactobacillus* including *L. reuteri.* Low concentrations of *L. casei* yield maximal levels of IL-6, IL-12, and TNF- α but virtually no IL-10 induction. At higher concentrations, IL-10 production increases radically with no change in the level of the other cytokines. This indicates, species of *Lactobacillus* exert very different and even opposing effects on DC activation. The DC polarizing factors such as LPS and bacterial DNA that stimulate IL-12 production in DC favors type 1 response. In contrast, factors suppressing IL-12 production favors type 2 response. Whereas IL-10 has been suggested to be a type 3 factor, because it not only suppresses IL-12 but also attenuates the stimulatory capacity of the DC. It might be speculated that *L. casei*, as a strong IL-12 and TNF- α inducer as well as MHC class II and B7-2 inducer, could be a contributing type 1 DC-polarizing factor when present in appropriate amounts in the gut, whereas *L. reuteri*, due to the DC-stimulating capacity characterized by only a weak MHC class II and B7-2 induction and suppressed IL-12 induction, could be a type 2 or 3 DC-polarizing bacteria. This indicates that the threshold for the bacterial concentration necessary to induce cytokine production varies among cytokines. Accordingly, individual variations in the concentration of bacteria may, in addition to the presence of different types of bacteria, generate different outcomes following interaction with local DC and thus ultimately dictate whether a Th1, Th2, or Th3 response occurs.

Several probiotic bacteria produce a variety of antimicrobial compounds (e.g., short-chain fatty acids, hydrogen peroxide, nitric oxide, bacteriocins) that may enhance their ability to compete against other GI microbes and which could potentially inhibit pathogenic bacteria. Traditionally, bacteriocin production has been an important criterion in the selection of a probiotic strain. Bacteriocins are diverse group of ribosomally synthesized antimicrobial peptides produced by bacteria. Bacteriocins produced by Lactic Acid Bacteria are classified into three main catagories:

Class I lantibiotics

Type A: Elongated shaped molecules

Type B: Globular molecules

Class II small heat-stable non lantibiotics

Subclass IIa: Pediocin-like bacteriocins

Subclass IIb: Two-peptide bacteriocins

Subclass IIc: Other peptide bacteriocins

Class III large heat-labile proteins.

Analysis of the known genomic sequences of *Lactobacillus* strains including *L. plantarum, L. acidophilus NCFM, L. johnsonii NCC 533,* and *L. sakei* predicts a broad group of bacteriocins with divergent sequences. These peptides have a relatively narrow spectrum of activity and are mostly toxic to Gram-positive bacteria, including *Lactococcus*, *Streptococcus*, *Staphylococcus*, *Listeria*, and *Mycobacteria.* The primary mechanism of bacteriocin action involves the formation of pores in the cytoplasmic membrane of sensitive bacteria and sometimes may interfere with the essential enzyme activity of sensitive strains. Formation of pores in the cytoplasmic membrane has dissipation of proton motive force (pmf), ATP depletion and leakage of nutrients and metabolites. Though pore formation is a common feature of bacteriocin, the size, stability, and conductivity of theses pores varies with respect to the type of bacteriocin.

Bacteriocins contribute to probiotic potentiality of a producer in many ways. Bacteriocins act as colonizing peptides, promoting the introduction and/or dominance of a producer into an already occupied niche. They function as killing peptides, directly forbidding competing strains. Bacteriocins sometimes function as signaling peptides, either signaling other bacteria through quorum sensing and bacterial cross talk within microbial communities or signaling cells of the host immune system. GI tract harbors high density of micro organisms resulting in close cell to cell interaction within the species and between the species, encouraging both mutualism and antagonism. This enables the producer to be a potential competitor over the sensitive strains in the niche. Many studies demonstrated the ability of bacteriocin producing lactobacilli inhibiting pathogens in the GI tract. *Lactobacillus plantarum*DDEN11007 inhibits *L. monocytogenes* EP2, producing the pediocinAcH. A two-peptide bacteriocin Abp118 prodved by *L. salivarius*UCC118 also inhibits *L. monocytogenes*.Nisisn Z, a bacteriocin of *L. lactis* inhibits vancomycin resistant enterococci, (For more details refer chapter 9).

Quorum sensing promotes bacterial population to synchronize group behavior and make coordinated multicellular functionality easier. Bacteriocins at higher concentrations function as inhibitors and at low concentrations serve as signaling molecules. Histidine protein kinase (HPK) located on the surface of the cell membrane and an intra cellularly located response regulator are the two important components of regulatory signal transduction system. HPK and RR are responsible for sensing of the signaling peptide and inducing an appropriate cellular response. But in case of auto inducing systems, when signaling peptide is produced above a certain threshold concentration, the autoinducing peptide binds to the N-terminal domain of the HPK, resulting in auto phosphorylation and activation. HPK then transfers a phosphoryl group to the RR, which ultimately give rise to a response, often at the level of transcription. Usually, auto inducing peptide has no function other than signaling, however, rarely function as antimicrobials. Nisin is an example of bacteriocin with both signaling and antimicrobial property.

Organic acids produced by lactobacilli serve as inhibitory metabolites, confining the colonization of competitive strains. Lactic, propionic and acetic acids are group of organic acids produced by lactobacilli benefits the producer, lowering the pH in the colonized region, inhibiting the growth of wide range of Gram negative pathogenic bacteria. As discussed in the previous chapters, Short-chain fatty acids (SCFA) such as acetic, propionic, and butyric acid, which are the fermentation products of saccharides, are often found in the colon of animals at various concentrations. Their presence in the human colon can positively alter the gut physiology drastically. SCFAs affect important biological processes, such as growth, metabolism, and differentiation of the intestinal epithelial cells (IECs); these processes are vital in maintaining the intestinal barrier integrity. Acetic acid is the chief SCFA produced in the colon. Once it enters systemic circulation, it serves as the primary source for lipogenesis by acetyl-CoA synthetase in the cytosol of adipose tissue. Propionic acid, a substrate for gluconeogenesis inhibits cholesterol synthesis in hepatic tissue. Butyric acid is a principal metabolic fuel for IECs and influences their proliferation and differentiation.

Understanding the ecology of *Lactobacillus* in the gastrointestinal tract gives insight into the mechanism of action of probiotic properties of the genus. However, the mechanism by which *Lactobacillus* spp. exert its beneficial influence on the host is still remained incomplete. Studies of the components that help *Lactobacillus* to make its space in the colon, competing with other bacteria, challenging complex host environment invigorate the probiotic research. Advances in modeling and analysis of beneficial aspects of lactobacilli would enhance the knowledge of their role in maintaining good health of its host, allowing customization of existing and future therapeutic and prophylactic modalities.

Key questions

1. Which of the following statement regarding Horizontal gene transfer (HGT) is not correct.
   a) A significant mechanism responsible for exchange of bacterial genes that leads to the acquisition of new characters.
   b) It is the process by which genetic information, including mutations, is passed from a parent to its offspring.

c) Fundamental lateral gene transfer process for the evolution of microbial genome.

d) Genes responsible for biosynthesis contribute to the metabolic adaptation through HGT.

2. Role tube method for isolation of novel anaerobic bacteria particularly from gastrointestinal tract was developed by........................

a) Frederick Griffith

b) Stanley Cohen and Herbert Boyer

c) Hungate

d) Morgan

3. MME stands for

a) Molecular Microbial Ecology

b) Metabolic Microbial Ecology

c) Medical Microbial Ecology

d) Mendelian Microbial Ecology

4. Write a detailed account on the mechanism of interaction lactobacilli with other microorganisms in the GIT.

Further reading

Barbieri, F., Montanari, C., Gardini, F., & Tabanelli, G. (2019). Biogenic amine production by lactic acid bacteria: A review. Foods, 8(1), 17.

Bintsis, T. (2018). Lactic acid bacteria as starter cultures: An update in their metabolism and genetics. AIMS microbiology, 4(4), 665.

Bintsis, T. (2018). Lactic acid bacteria as starter cultures: An update in their metabolism and genetics. AIMS microbiology, 4(4), 665.

Feng, T., & Liu, Y. (2022). Microorganisms in the reproductive system and probiotic's regulatory effects on reproductive health. Computational and structural biotechnology journal, 20, 1541-1553.

Garcia-Gonzalez, N., Prete, R., Battista, N., & Corsetti, A. (2018). Adhesion properties of food-associated Lactobacillus plantarum strains on human intestinal epithelial cells and modulation of IL-8 release. Frontiers in microbiology, 9, 2392.

Gharbi, Y., Fhoula, I., Ruas-Madiedo, P., Afef, N., Boudabous, A., Gueimonde, M., & Ouzari, H. I. (2019). In-vitro characterization of potentially probiotic Lactobacillus strains isolated from human microbiota: interaction with pathogenic bacteria and the enteric cell line HT29. Annals of Microbiology, 69, 61-72.

Hatti-Kaul, R., Chen, L., Dishisha, T., & Enshasy, H. E. (2018). Lactic acid bacteria: from starter cultures to producers of chemicals. FEMS Microbiology Letters, 365(20), fny213.

He, Y., Niu, X., Wang, B., Na, R., Xiao, B., & Yang, H. (2020). Evaluation of the inhibitory effects of Lactobacillus gasseri and Lactobacillus crispatus on the adhesion of seven common lower genital tract infection-causing pathogens to vaginal epithelial cells. Frontiers in medicine, 7, 284.

Kayama, H., Okumura, R., & Takeda, K. (2020). Interaction between the microbiota, epithelia, and immune cells in the intestine. Annual review of immunology, 38(1), 23-48.

Liu, Q., Yu, Z., Tian, F., Zhao, J., Zhang, H., Zhai, Q., & Chen, W. (2020). Surface components and metabolites of probiotics for regulation of intestinal epithelial barrier. Microbial Cell Factories, 19, 1-11.

Nishiyama, K., Sugiyama, M., & Mukai, T. (2016). Adhesion properties of lactic acid bacteria on intestinal mucin. Microorganisms, 4(3), 34.

Petrova, M. I., Reid, G., Vaneechoutte, M., & Lebeer, S. (2017). Lactobacillus iners: friend or foe?. Trends in microbiology, 25(3), 182-191.

Popov, D. V. (2022). Microbiota and reproduction in agricultural mammals. Agricultural Biology, 57(2), 222-236.

Von Wright, A., & Axelsson, L. (2019). Lactic acid bacteria: an introduction. In Lactic acid bacteria (pp. 1-16). CRC Press.

Wang, J., Ji, H., Wang, S., Liu, H., Zhang, W., Zhang, D., & Wang, Y. (2018). Probiotic Lactobacillus plantarum promotes intestinal barrier function by strengthening the epithelium and modulating gut microbiota. Frontiers in microbiology, 9, 1953.

Wang, Y., Wu, J., Lv, M., Shao, Z., Hungwe, M., Wang, J., ... & Geng, W. (2021). Metabolism characteristics of lactic acid bacteria and the expanding applications in food industry. Frontiers in bioengineering and biotechnology, 9, 612285.

Wu, Y., Zhu, C., Chen, Z., Chen, Z., Zhang, W., Ma, X., ... & Jiang, Z. (2016). Protective effects of Lactobacillus plantarum on epithelial barrier disruption caused by enterotoxigenic Escherichia coli in intestinal porcine epithelial cells. Veterinary immunology and immunopathology, 172, 55-63.

Yu, Q., Yuan, L., Deng, J., & Yang, Q. (2015). Lactobacillus protects the integrity of intestinal epithelial barrier damaged by pathogenic bacteria. Frontiers in cellular and infection microbiology, 5, 26.

5

Exopolysaccharide Production by Intestinal Lactobacilli

Introduction

Cell surface components such as adhesins, polysaccharides, and proteins play major roles in the adherence of probiotic bacteria to the intestinal epithelium, interactions that might lead to pathogen exclusion and immunomodulation of host cells. Probiotic bacteria have the ability to produce exocellular polymers called Exopolysachcharides (EPSs). These exocellular polymers present in the surface of many lactic acid bacteria, including Lactobacilli.The adhesive properties of lactobacilli are directly linked to their surface properties which are influenced by the structure and composition of their cell wall. The cell envelope of lactobacilli, bacteria, is composed of the bilipidic plasma membrane with embedded proteins encompassed by the cell wall. The bacterial cell wall consists of a thick multilayered sacculus made of peptidoglycan (PG) with teichoic acids (wall teichoic acids (WTA) and/or lipoteichoic acids (LTA)), exopolysaccharides (EPS), proteinaceous filaments called pili, and proteins that are anchored to the cell wall through different mechanisms (Figure 1).

Some species of lactobacilli possess an additional paracrystalline layer of proteins surrounding the PG layer, referred to as the S layer. These macromolecules together may play crucial roles in determining species and strain-specific characteristics of lactobacilli by influencing host-microbe interactions and microbial adaptations to the changing host environment. Several studies implicate that cell surface components, either individually or collectively, involve in host-microbe interactions. Lactobacilli beneficially affect the host in many ways by producing abundant variety of EPSs through the activity of extracellular glycosyl transferases. Approximately 30 species of lactobacilli are described as EPS producers. Among them, the best known are *L. casei, L. acidophilus, L.brevis, L. curvatus, L. delbrueckii bulgaricus, L. helveticus, L. rhamnosus, L. plantarum, L. johnsonii, etc.*

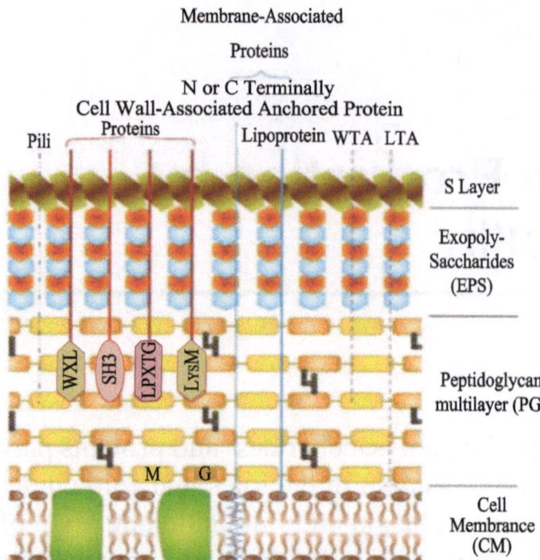

Fig. 1. Schematic representation of cell envelope of lactobacilli

Characteristics of the EPS

EPS act as prebiotic substrates and contribute to host health, promoting the growth of other probiotic bacteria including Bifidobacteria. EPS produced by lactobacilli have several medicinal importance possessing antitumor, antimutagenic, antiulcer, antibacterial, immunomodulating and blood cholesterol lowering activities. EPS being the major component of biofilms, play a key role in protecting the producer organism against phagocytosis, phage attacks, lysozyme, predatory attack from protozoa, dessication, antibiotics, toxic compounds (e.g. toxic metal ions, sulfur dioxide, and ethanol), osmotic stress and bacteriocins such as nisin. In addition, EPSs retain longer in the GI tract and enhances the colonization of its producer. EPSs enable bacteria to survive in the harmful environmental conditions such as too low or high pH, gastric acid and bile salts.

Biosynthesis of EPS and its Types

EPS is secreted mainly in two forms: (a) as a capsular exopolysaccharide which is associated with the cell surface (b) as slime exopolysaccharide secreted as free polymers to the environment. EPS from LAB are high molecular mass sugar polymers encompassing two main groups namely: homopolysaccharides (HoPS) and heteropolysaccharides (HePS).

Homo exopolysaccharides (HoPS)

HoPS, consist of single type of monosaccharide (α-D-glucans, β-D-glucans, fructans, and others represented by polygalactan) and HePS, composed of different types of monosaccharides, mainly D-glucose, D-galactose, L-rhamnose, and their derivatives. Two important groups of HoPSs produced by LAB includes: (i) α-glucans, mainly composed of α-1,6- and α-1,3-linked glucose residues, namely dextrans (ii) fructans, mainly composed of β-2,6-linked fructose molecules, such as levan. Generally, HoPSs molecular weights range between 4.0 $\times 10^4$ and 6.0 $\times 10^6$ Da. These are produced at higher concentrations than HePSs.

Glucans are composed of a glucose polymer backbone with various degrees of branching depending on producing strains. For instance, dextrans from *Lactobacillus fermentum, L. sakei, L. parabuchneri* and *L. hilgardi* are α-(1,6) glucans ramified by glucose residues at position 3 and less frequently at positions 2 and 4. In *Lactobacillus* species, namely, *L. johnsonii, L. reuteri, L.sakei, L. fermentum, L. parabuchneri* and *L. sanfranciscensis* reuteran sucrases, glucan-type HoPS, as well as a levansucrases and inulosucrases, which synthesize fructan-like HoPS, have been characterized to date. Subspecies of *L. reuteri* are able to generate other types of EPS such as levan, inulin-type fructan, mutan, reuteran in a large range of molecular weights. Mutan, a glucan that differ from dextran in containing high percentage of α-1, 3 linkages. Glucans with α-1,6 are readily soluble in water while, glucans with α-1, 3 linkages are less soluble. Compared to soluble glucans, insoluble glucans help in adherence to the mucosal surface. Alternan, another glucan with alternate α-1,6 and α-1,3 linkages, and this structure is thought to be responsible for its distinctive physical properties including high solubility and low viscosity. In addition, levan is a fructan, produced from sucrose, composed of β-2,6-linked fructose molecules with some β-2,1-linked branches.

Homo EPS are synthesized outside the cell by specific glucosyltransferase (GTF) or fructosyltransferase (FTF) enzymes (commonly named glucansucrases or fructan-sucrases). GTFs of LAB composed of four domains namely: (i) a N-terminal signal peptide, (ii) a N-terminal variable region, (iii) a catalytic core, and (iv) the C-terminal. Catalytic core is highly conserved and C-terminal includes glucan-binding region. Similarly, FTFs also possess four different regions: (i) N-terminal signal peptide directing protein export, (ii) N-terminal variable region, (iii) the highly preserved catalytic core region of 450 to 550 aminoacids, (iv) the C-terminal sometimes containing a cell wall binding domain, helps to anchor the producing organism. A characteristic feature of FTFs of lactobacilli is the presence of direct repeats in the N-terminal variable region, which are typically absent in the FTFs of other bacteria.Energy from sucrose hydrolysis is sufficient for HoPS synthesis. The FTFs catalytic region contains amino acid motifs that are highly conserved in all bacterial levansucrases and other fructan active enzymes. The C-terminal anchor region consists of several PXX repeats that span the cell wall and a LPXTG motif directing covalent linkage to the peptidoglycan.HoPS are involved in resistance to certain stress conditions. Accordingly, FTFs are found to be over expressed in the presence of environmental stress that affects biophysical properties of the cytoplasmic membrane.

There is no much energy is required for EPS-production other than for enzyme biosynthesis because EPS synthesis by GTF or FTF does not involve active transport processes or the use of activated carbohydrate precursors. Therefore, large amounts of sucrose can easily be converted to EPS. GTFs use energy of the osidic bond of sucrose to catalyze the transfer of glycosyl moiety of sucrose to H_2O, an acceptor carbohydrate, or a growing polymer chain. The reactions for biosynthesis of fructans and oligo fructanscatalyzed by FTFs can be written as follows:

sucrose+ H_2O ⟶ fructose +glucose
sucrose+ acceptor carbohydrate (C) ⟶ oligosaccharide (CF)+glucose
sucrose +fructan (Gfn) ⟶ fructan(Gfn+1)+glucose

The net reaction for biosynthesis of glucan and gluco oligosaccharides catalyzed by GTFs can be written as follows:

Sucrose+ H_2O ⟶ glucose+fructose
Sucrose+ acceptor carbohydrate (C) ⟶ oligosaccharide (CG) +fructose
Sucrose + glucan (FGn) ⟶ glucan (FGn+1) + fructose

Although the net reactions catalyzed by GTFs are similar to FTFs, the molecular mechanisms catalyzed by GTFs are related to glucan degrading enzymes such as amylases rather than FTFs. GTFs and FTFs play a key role in sucrose metabolism. GTFs and hydrolase activity of FTFs liberate fructose from sucrose that further reduced to mannitol with concomitant oxidation of NADH to NAD^+, enabling the formation of acetate from acetyl phosphate and the gain of an additional mole of ATP in the metabolic flux through the pentose phosphate pathway.

Hetero exopolysaccharides

HePS are polymers of a repeating unit (anoligosaccharide of three up to eight monosaccharides) that is synthesized out of sugarnucleotides intracellularly and polymerized extracellularly. Many enzymes are involved in their synthesis and secretion. Chemical composition of HePSs show wide variability and are composed of at least two different sugars out of glucose, galactose and rhamnose at different ratios and their molecular weights vary from 10^4 Da to 6×10^6 Da. HePS are made by polymerization of repeating units, and these repeating units are built by a series of addition of sugar nucleotides at the cytoplasmic membrane.LAB strains can utilize various monosaccharides and disaccharides as energy sources, via some well-studied sugar uptake systems include primary transport systems, direct coupling of sugar translocation to ATP hydrolysis via a transport-specific ATPase; secondary sugar transport systems, coupling of sugar transport to the transport of ions or other solutes, both as symport and antiport transport systems; and group translocation systems, coupling of sugar transport to phosphorylation via the phosphoenolpyruvate (PEP)- dependent phosphotransferase system. *Lb. delbrueckii* subsp. *Bulgaricus is* generally galactose-negative and takes up lactose *via* a lactose/ galactose antiport transport system. The glucose moiety of imported lactose undergoes fermentation, while the galactose moiety is excreted via the lactose/ galactose antiport system. After the addition of a HePS repeating unit, the unit is exported through the cell membrane and becomes polymerized into the final HePS. Therefore, biosynthesis and secretion of HePS requires many enzymes and proteins, which may not be species specific. Once the sugars enter into the cell, they transformed to sugar nucleotides. Intracellular monosaccharides are converted to sugar nucleotide substrates for glycosylpyrophosphorylases catalyzed polymerization reactions including UDP (uridine diphosphate), dNTP (thymidine diphosphate), and GDP (guanosine diphosphate). Glycosidic linkages are formed on the membranes of cytoplasm. Then by priming glycosyl transferases, a sugar moiety is transferred to C55 polyprenyl phosphate, a carricr lipid and component of the membrane. This transfer triggers the addition of more repeating units to the HePS.The repeating unit is exported through the bacterial membrane, and is polymerized to become a hetero-EPS. Disruption of the gene for priming glycosyl transferases leads to the generation of non EPS producing mutants. Hence, glycosyl transferases are thought to be crucial for synthesis of HePS. C55 polyprenyl phosphate is also a part of cell wall biosynthesis and therefore, both EPS and cell wall biosynthesis compete for this substrate.

Depending on the type of producing species and environmental factors, the composition, structure and function of EPSs varies greatly. Studies suggest that biosynthesis of EPSs is under the control of quorum sensing through the regulation of gene expression of proteins necessary for the EPSs synthesis. Genes for EPSs synthesis may be located in plasmids as in case of *L. casei* or may present in chromosome as in case of thermophilic LAB.

As discussed above, polymers from lactobacilli mainly include levan- and inulin-type fructans and the a-glucans dextran, mutan and reuteran. The sucrose supplementation in human diet is likely to support poly and oligosaccharide synthesis by glycosyltransferases of lactobacilli in the upper intestine. Fructan and glucan production are generally associated with the formation of fructo-oligosaccharides (FOS) and glucose oligosaccharides respectively. In lactobacilli, homopolysaccharide and oligosaccharide production is most frequently found in those species that are associated with gastrointestinal ecosystems, for example, *Lactobacillus acidophilus, Lactobacillus johnsonii, Lactobacillus mucosae* and *Lactobacillus reuteri*. Fructan and levan type EPS produced from *L. sanfranciscensis* enhance bifidogenic effect in the human gut. EPSs from *Lactobacillus* have shown to involve in thecell of adhesion. EPSs contribute to the physicochemical properties of the cell surface by shielding other cell surface adhesins, acting as ligands mediating adhesion and coaggregation. EPSs from strains of *L. plantarum* bind and inactivate mutagens like heterocyclic amines. EPS produced by *L. helveticus* ssp. *Jugurti* exhibit antitumor activity.

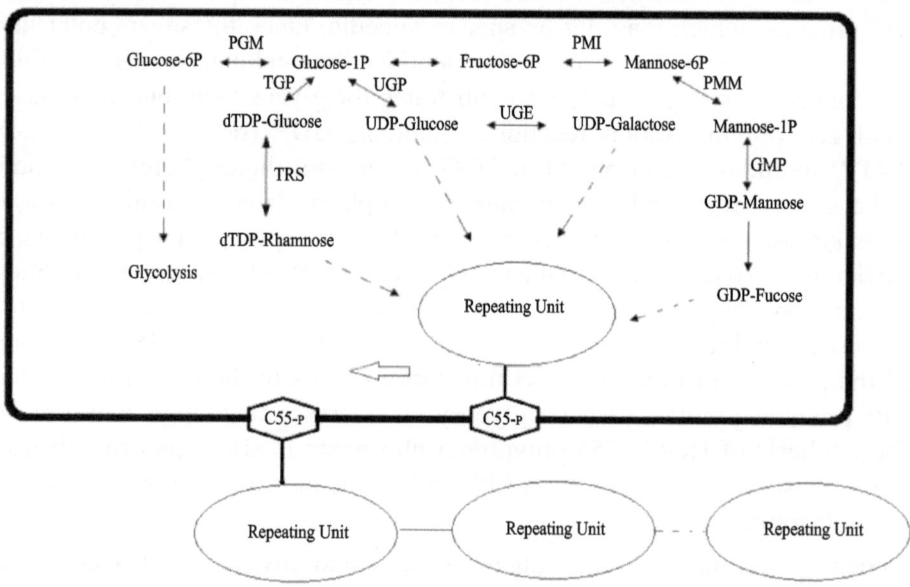

Outline of biosynthesis of HePS.

PGM: α-phosphoglucomutase; UGP: UDP-glucose pyrophospholyrase; UGE: UDP-galactose 4-epimerase; TGP: dTDP-glucose pyrophospholyrase; TRS: dTDP-rhamnose synthetic enzyme system; PMI: phosphomannoisomerase; PMM: phosphomannomutase; GMP: GDP-mannose pyrophospholyrase.

Key questions

1. Mechanism of biosynthesis of exopolysaccharide.
2. Detailed functions of enzymes involved in biosynthesis of hetero polysaccharides.
3. Glycosyltranferase and fructosyltransferase are the two enzymes involved in the
   a) Synthesis of hetero EPS outside the cells.
   b) Synthesis of hetero EPS inside the cells.
   c) Synthesis of homo EPS outside the cells.
   d) Synthesis of homo EPS inside the cells.
4. Explain the mechanism of adhesion of lactobacilli to the gastrointestinal cell surface.

Further readings

Bajpai, V. K., Rather, I. A., Majumder, R., Shukla, S., Aeron, A., Kim, K., ... & Park, Y. H. (2016). Exopolysaccharide and lactic acid bacteria: Perception, functionality and prospects. Bangladesh Journal of Pharmacology, 11(1), 1-23.

Bibi, A., Xiong, Y., Rajoka, M. S. R., Mehwish, H. M., Radicetti, E., Umair, M., ... & Aadil, R. M. (2021). Recent advances in the production of exopolysaccharide (EPS) from Lactobacillus spp. and its application in the food industry: A review. Sustainability, 13(22), 12429.

Eze, C. O., Berebon, D. P., Gugu, T. H., Anazodo, F. I., & Okorie, J. E. (2022). Lactobacillus exopolysaccharide: An Untapped Biopolymer. In Lactobacillus-A Multifunctional Genus. IntechOpen.

Jurášková, D., Ribeiro, S. C., & Silva, C. C. (2022). Exopolysaccharides produced by lactic acid bacteria: from biosynthesis to health-promoting properties. Foods, 11(2), 156.

Nachtigall, C., Surber, G., Herbi, F., Wefers, D., Jaros, D., & Rohm, H. (2020). Production and molecular structure of heteropolysaccharides from two lactic acid bacteria. Carbohydrate polymers, 236, 116019.

Rana, S., & Upadhyay, L. S. B. (2020). Microbial exopolysaccharides: Synthesis pathways, types and their commercial applications. International journal of biological macromolecules, 157, 577-583.

Sajna, K. V., Sharma, S., & Nadda, A. K. (2021). Microbial exopolysaccharides: An introduction. Microbial exopolysaccharides as novel and significant biomaterials, 1-18.

6

Uses of Probiotics in Controlling Various Types of Microbial Infections; Bacterial Therapeutics

Introduction

It became clear from the discussions of previous chapters that probiotics perform metabolic functions such as fermenting indigestible dietary residues and endogenous mucus, production of vitamins, and absorption of minerals. They also play key role in intestinal epithelial cell proliferation and differentiation, and the homeostasis of the immune system. Further, probiotics are also known to involve in modifying gut pH, antagonizing pathogens through the production of antimicrobial compounds, competing with pathogens for recept or sites as well as for available nutrients and growth factors, stimulating immunomodulatory cells of the host, and produce a wide range of enzymes such as lactase, amylase, lipase, protease, phytase and so on. The molecular mechanisms of probiotics are largely unknown and opens avenue for research in probiotics. Moreover, current research in probiotics and health consciousness in the public lead to an increased interest towards probiotics from the public, researchers, governmental organizations (such as the WHO/FAO) and medicine and food companies. In this chapter, application of probiotics in treating various diseases and disorders are discussed.

Probiotics for infectious and antibiotic associated diarrheal diseases

Diarrhea is simply an altered movement of ions and water that follows an osmotic gradient. Enteric infections mediated diarrhea is more prevalent in children, accounting for high morbidity and mortality worldwide with an estimate of 2-4 billion episodes every year. Infectious diarrhea has known to associate mainly with three classes of pathogens namely bacteria, viruses and parasites. Major bacterial diarrheal pathogens include *Vibrio cholerae, Clostridium difficile, Shigella spp. and E.coli spp.* while, rotavirus, norovirus and astrovirus infections results in viral diarrhea whereas, *Giardia lamblia* and

*Entamoeba histolytica*are parasitic diarrheal pathogens. In addition to the infectious diarrhea, Antibiotic associated Diarrhea (AAD) is also one of the most common gastrointestinal side effectof antibiotic therapy often associated with *C. difficile* infections in adults and children.

Infection by enteric pathogens alters intestinal microbial composition that changes carbohydrate metabolism with decreased short chain fatty acid absorption and loosening the tight junctions ultimately leads to osmotic diarrhea. Probiotics proved to control infectious diarrhea and efficiently manage the consequences raised as a result of antibiotic treatment. Probiotics help host to fight diarrhea, competing with the pathogens for nutrition, host cell surface receptors for binding, producing bacteriocin like antimicrobial compounds, modulating host immunity and balance intestinal microbiota.

One of the major problem associated with antibiotic use, particularly clindamycin, cephalosporins, and penicillins, is the onset ofdiarrhea caused by *Clostridium difficile*. Disruption of intestinal microbiota by antibiotics leads to an abnormal elevation of *C. difficile* with a significant decrease in the number of beneficial bacteria.Followed by an antibiotic treatment, administration of probiotic functional foods help in reestablishing the diminished microbiota enriched with beneficial bacteria outnumbering the harmful microorganisms. Administration of *Lactobacillus GG* decreases stool frequency and increases stool consistency during antibiotic therapy and has been proven successful in treating *C. difficile* relapsing diarrhea, without any side effects, in both children and adults. In addition, strains of *Lactobacillus reuteri, Streptococcus boulardii, and Bifidobacteria* spp., were also found to be beneficial in curing both AAD and infectious diarrhea.

Role of probiotics in inhibiting rotaviral diseases

Further, probiotics are also able to suppress travelers' diarrhea and other diarrheal diseases in young children caused by rotaviruses. The probiotic species generally used for children include *L. reuteri, L. casei, S. boulardii, B. bifidum* and *S. thermophilus*. In this chapter, mechanism of action of probiotics against viral diarrhea particularly rotaviral diarrhea has discussed in detail. Many theories were postulated to demonstrate the potential mechanisms of probiotics in reducing the duration of rotaviral diarrhea. But these theories have many flaws.The first postulated theory explained that lactobacilli competitively binds and blocks the receptor sites on intestinal epithelial cells and thus prevents adhesion and invasion of the virus. Due to insufficient evidence the concept seem to be plausible. To explain, in most of the cases, by the time of probiotic administration, patients had suffered diarrhea for possibly 12h. Interim, the virus by taking upper hand has infected mature enterocytes in the mid- and upper region of the small intestinal villi further increasing their number and spread rapidly. By then, virus and /or its enterotoxin will have inhibited electrolyte transport and also thereby lowering fluid and glucose absorption further worsening the condition. In addition, secretory reflexes in epithelia being potentially activated by viral enterotoxin cause loss of fluids resulting in diarrhea. In such a condition, if ingested probiotic lactobacilli somehow compete with the toxins or peptides released from infected or altered villous endocrine cells, then it is possible to arrest the cascade of physiological reactions leading to diarrhea.

Probiotic mediated enhancement of the immune response against virus has thought to be the second mechanism to prevent rotaviral diarrhea. Attenuation of virus and /or effect of its toxin on enteric nervous system, alteration in the cytokine profile with an increase in anti-inflammatory effect might provide rapid cessation of epithelial secretion and diarrhea. Alternatively, LAB proven to stimulate the production of gamma interferons by T cells leading to potential inhibition of chloride secretion this effect might also inhibit diarrhea.

Removal of immune recognized noxious substances by probiotic signal mediated down regulation of secretary and motility defenses has thought to be the third mechanism of probiotics against rotaviral diarrhea. To explain further, glycosylated intestinal mucins inhibit rotavirus. In the intestine, signals from probiotics increase the expression of *MUC2* and *MUC3* mucin encoding genes and confers protection to host against pathogen adhesion. Final theory proposed that probiotics particularly LAB produce metabolic compounds that inactivate viral particles.However, the more specific antiviral properties have not been ruled out, the effect may be due to the production of acid and anti viral amino acids like cystine. The theories discussed are the possible ways in reducing the duration of rotaviral diarrhea using probitics. Further detailed research is required to understand how probiotic strains reduce the duration of diarrhea in conjunction with rehydration therapy.

Helicobacter pylori infections

Helicobacter pylori is a gram negative, micro aerophilic and acid sensitive bacteria known to cause gastritis, peptic ulcers and gastric cancers. It is worldwide in distribution with a prevalence of 30-60%. In order to survive in acidic condition, *H. pylori* relies on its urease enzyme that increases the pH of its surrounding environment by hydrolyzing urea into carbon dioxide and ammonia. Lactic Acid Bacteria proven to impede the growth of *H. pylori* and make its survival very difficult by inhibiting urease activity. In addition, probiotic metabolites namely Bacteriocins and SCFAs also play an important role in controlling the establishment of *H. pylori*in human. Further,

Lactobacillus spp., Enterococccus faecium, Bacillus subtilis and Bifidobacteria produce bacteriocins which are capable of inhibiting antibiotic -sensitive and antibiotic-resistant strains of *H. pylori.* To mention, Nisin A, pediocin PO2, leucocin K, lacticin A164 and lacticin BH5 are some of the LAB bacteriocins with anti *H. pylori* activity. SCFAs such as formic, acetic, propionic, butyric and lactic acids create an acidic environment and inhibit the pathogen. In addition, by suppressing the expression of inflammatory mediators such as TNF α, IL- 8, inducible nitric oxide synthase and cyclooxygenase- 2, with an concurrent enhancement in the production of anti inflammatory immune compounds probiotics relieves inflammation in *H. pylori* infected individual. Therefore, the use of probiotics to treat *H. pylori* infection has been proposed for improving eradication rate and tolerability.

Probiotics for preventing colon cancer

Among various types of cancers, colorectal cancer represents a major public health problem accounting for over 1 million cases and about half a million deaths worldwide. Survival from colon cancer has been found to vary demographically and estimated to be 65% in North America, 54% in Western Europe, 34% in Eastern Europe, and 30% in India. Although chemotherapy and radiotherapy have been applied as the surgical adjuvant treatments of colon cancer, they vary in success rates for local recurrence, disease-free survival, and overall survival. In addition, these treatments may present some side effects such as an increased risk for infections, hair loss, fatigue, lip sores, nausea, vomiting, diarrhea, and bloody stools. Diet interventions and natural bioactive supplements have now been extensively studied to reduce the risks of colon cancer, as a cause of prevention instead of cure. The use of probiotics to prevent colon cancer has gain much attention due to the positive results from *in vivo* studies. Postulated probiotic mechanisms to prevent colon cancer include reduction in the activity of several cancer causing bacteria, modification of differentiation processes in tumor cells, production of short chain fatty acids, alteration of tumor gene expressions, desmutagenic and anticarcinogenic properties. In addition, prebiotics being indigestible, have also been associated with improved bowel functions and metabolisms of the distal colon, including a reduced risk of colon cancer.

Following are the different strategies of probiotics for prevention and management of colon cancer:

Pathogen exclusion: Several microbial infections are known to associate with the development of cancer and it has been reported that 17.8% of global cancer burden is due to infections. Further, in developing countries the rate of infection associated cancer is 26.3% while in developed countries, 7.7%.

Therefore, elimination of pathogens would help in preventing cancer and thereby reduce the burden of infection associated cancer in humans. Probiotics have proven to exclude pathogens by their mechanism of action and found useful in preventing infections. Immunomodulatory properties of probiotics such as increased immune response by elevated levels of sIgA and anti inflammatory cytokines, production of antiviral and antimicrobial compounds, SCFAs, that aids in restricting the pathogenesis and thereby confers protection against infection associated colon cancer.

Anti mutagenic property of probiotics: Development of cancer is a multi stage process that would initiate when mutations accumulate in the tumor suppressor and protooncogenes. The mutations may activate oncogenes or inactivate tumor suppressor genes or sometimes both. Probiotics, particularly lactobacilli proven to reduce mutagen induced chromosome aberrations by 80% *in vivo*. The possible mechanism that confers antimutagenic or anticarcinogenic property involves the binding of mutagenic compounds such as heterocyclic amines by the probiotic bacteria, reducing their absorption in intestine, minimizing their retention time and eliminating through feces. Interestingly, live probiotics bound or inhibit the mutagens permanently, compared to killed bacteria indicating that the interaction with mutagens may influence the bacteria at genetic level to express the formation of some of the components, may be cell wall proteins to bind firmly or to inhibit by transforming the mutagens.

Inhibition of xenobiotic enzymes: The products and intermediates of xenobiotic metabolism also take part in the development of carcinogenesis. The enzymes responsible for the xenobiotic metabolism are generally referred as xenobiotic metabolic enzymes or xenobioitc enzymes. Intestinal pathogens produce these enzymes at higher levels which catalyze an array of reactions resulting in the formation of xenobiotic derived or transformed genotoxic and carcinogenic compounds. To mention, putrefactive bacteria mainly *Eubacterium spp., Clostridium clostridiiforme, C. paraputrificum, C. nexile, Butyrivibrio spp., and Bacteroides spp.* produce higher levels of enzymes such as β-glucuronidase, nitroreductase, nitrate reductase and azoreductase that tend to deconjugate xenobiotic conjugates and release metabolites such as aglycones and other procarcinogenic substances in the intestine which further undergo reabsorption and recirculation thus extending their retention time in the body they greatly increase the risk of cancer. Probiotics by inhibiting the activity of xenobiotic enzymes confers protection to the host against xenobiotic induced carcinogenesis. Further, as putrefactive bacteria produce xenobiotic enzymes at higher levels, use of probiotics with appropriate prebiotics also found to be more effective in restricting them by promoting saccharolytic species.

Production of protective metabolites

Probiotics produce a wide range of metabolites that confer health benefits to the host. The metabolites of probiotic bacteria such as ariginine, glutamine, SCFAs, bacteriocins and hydrogen peroxide are protective to the intestine against cancer. SCFAs produced during prebiotic fermentation inhibit mutagens at appreciable levels. Further, metabolic end products of fermentation induce apoptosis features such as DNA laddering, chromatin condensation, cell cycle arrest, accumulation of reactive oxygen species, disruption of mitochondrial transmembrane potential, formation of apoptotic bodies, caspase activation and release of cytochrome C resulting in programmed cell death a strategy in controlling tumorigenesis. Among the fermented metabolites, butyrate has proven to be more effective in the prevention and management of cancer. Butyrate nourishes the colonic mucosa and promotes the apoptosis of transformed colonocytes. It arrests the cell growth by altering the transcription of small number of genes responsible for cell differentiation and induces apoptosis in tumour cells. However, unfortunately many of the probiotic metabolites with anti carcinogenic property are not yet characterized. Screening and characterization of these metabolites with the help of structural, functional and comparative genomics may help in exploring the significant properties for the betterment of human health. Once the expression of genes that code for these metabolites are located in the sequenced genome, production of metabolites can be enhanced, improving the strains of probiotics through biotechnological approach.

Bacterial therapeutics for the treatment and prevention of Urogenital infections

Urogenital Tract Infections (UTI) are one of the major diseases responsible for loss of quality of life in women. UTI are common in women of age 17-30 years and it has been estimated that several hundred million women suffer from UTI worldwide amounting $6 billion annually for urogenital tract health care products. Further, it has estimated that nearly one billion women around the world suffer from non sexually transmitted urogenital infections, including bacterial vaginosis and yeast vaginitis annually. Gram negative organisms, particularly *E coli* (up to 85%), *Enterococcus faecalis* and *Staphylococcus saprophyticus* are the predominant causative agents in most women suffering from UTI.In case of recurrent bacterial vaginosis, uropathogenic *E. coli* and anaerobic Gram negative rods generally *Gardnerella vaginalis* predominates. While in recurrent yeast vaginitis, *Candida albicans* predominates. Resident microbial composition plays an important role in maintaining the health of vagina. Healthy human vagina is slightly acidic due to predominant lactobacilli species as common microbiota (Figure 1). In addition other dominant vaginal microbiotia include *Bifidobacterium, Gardnerella, Gemella, Atopobium, Corynebacterium and Janthinobacterium.* Due to antibiotic therapy, if vaginal Lactobacilli number declines then it leads to an increase in the pH and favours the accumulation of other pathogens. The most common example for such over growth is *Candida albicans* which is generally present in small number in the vagina. Out numbering of such opportunistic pathogens leads to urogenital infections.

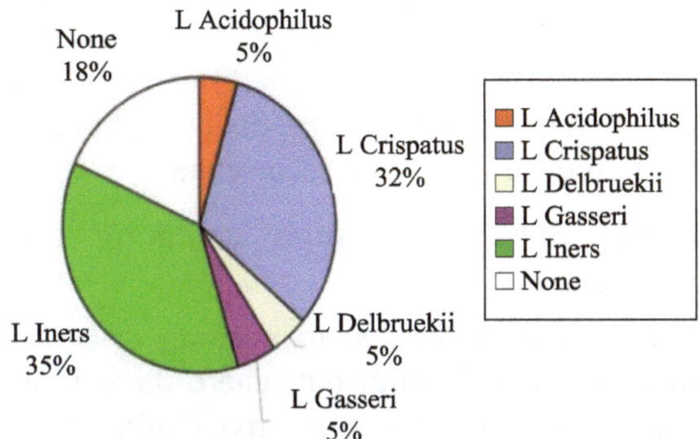

Fig. 1. The lactobacilli composition from vaginal sample

While treating, the primary aim is to restore vaginal indigenous lactobacilli and the maintenance of acidic pH that helps to protect against infection. Therefore, hydrogen peroxide producing probiotics can be used to treat UTI effectively. Combination of two strains of lactobacilli has recommended for vaginal use comprising *L. rhamnosus* GR-1a distal urethral isolate, selected primarily for its anti-Gram negative activities and resistance to spermicide, and *L. fermentum* RC-14, for anti-Gram positive cocci activities and hydrogen peroxide production. Studies showed that daily oral intake of probiotic strains *Lactobacillus rhamnosus* GR-1 and *Lactobacillus fermentum* RC-14, by patients with bacterial vaginosis resulted in the reversion of vaginal microbiota dominated with lactobacilli. It has thought that the ascention of indigenous lactobacilli or administered probiotics from colorectal region to vagina restores lactobacilli number and confers protection against pathogens. Another possible effect may be the enhanced intestinal mucosal immunity by probiotics administration rendering the urogenital tract environment less receptive to pathogens.

Further, yeast vaginitis is another common problem, estimated to affect around 1:5 black American women and close to 1:10 white women during any given two month time frame, with 1:12 reporting four or more episodes per year.

Use of broad spectrum antibiotics promote the over growth of fungal pathogens disrupting the normal vaginal mirobiota. *Candida albicans* is the major cause of yeast vaginitis (around 85%), while other yeast such as *Candida glabrata, Candida krusei*, and *Candida tropicalis* cause vaginitis comparatively to a lesser extent. A very few lactobacilli with anti fungal activity are able to kill or inhibit adhesion of yeast to vaginal cells. Therefore, it is advisable to patients on antibiotic treatment to take yoghurt or probiotic formulation with antifungal activity as a means to prevent yeast vaginitis. More recently, *L rhamnosus* GR-1 and *L fermentum* RC-14 starins have proven to possess such properties and found useful in treating yeast infection mediated inflammation in vagina. But still there is little evidence that probiotics can effectively cure a symptomatic yeast vaginitis. At best, reliable probiotic products might be an adjunct to drug therapy as a means to reduce the risk of recurrence.

Management of Irritable Bowel diseases through probiotic and prebiotic interventions:

Inflammatory bowel disease (IBD) is a term applied to a group of bowel disorders in which inflammation is a major feature, but where there is no proven evidence that infection is the causative agent. Ulcerative Colitis (UC), Crohn's disease and Pouchitis are the major entities of IBD. Crohn's disease is a chronic inflammatory disease of the intestine which flares up with periods of inactivity and can affect any part of the digestive tract, from mouth to the anus. The most common symptoms of Crohn's disease are abdominal pain and diarrhoea. There is no known cure for Crohn's disease and therefore treatment involves the induction and maintenance of remission, correcting malnutrition, addressing complications, and thereby improving the quality of life in patients. Further, Ulcerative Colitis by definition is continuous inflammation starting in the rectum and restricted to the colon. Both UC and Crohn's disease are considered as the two important subtypes of IBD with overlapping symptoms. However, they differ by the intestinal localization and features of the inflammation. To explain, inflammation in UC is continuous and affects mainly rectum and colon, while in Crohn's disease inflammation can occur anywhere in the gastrointestinal tract. In UC, microscopic lesions restricted to the mucosa layer can be seen while in Crohn's disease lesions affect full thickness of the intestinal wall. Pouchitis, another form of IBD is a non- specific inflammation of the ileal reservoir and is generally associated with UC as the most common complication of proctocolectomy with ileal pouch-anal anastomosis in patients with UC.

Most of the animal models that develop spontaneous or chemically induced colitis influence greatly by the intestinal microbiota. In human, compared to healthy individuals, IBD patients significantly differ in their intestinal microbial composition being outnumbered with potential pathogens. Alteration in the intestinal microbial composition by enhancing the colonization of beneficial bacteria with concurrent reduction in pathogen numbers found to be effective as an alternative therapy in treating IBD. Administration of probiotics strains of lactobacilli, bifidobacteria, and *Saccharomyces* can exert a beneficial effect in balancing the intestinal microbiota. Following are the few mechanisms of action of probiotics which are thought to inter relate with the treatment of IBD. (a) probiotics by exhibiting more affinity towards limited number of receptors present on the epithelium, compete with microbial pathogens and inhibit their adherence and translocation; (b) immunomodulation and/or stimulation of immune function of gut-associated lymphoid and epithelial cells by probiotics (anti-inflammatory immune compounds, production of secretary antibodies); (c) probiotics can antagonize pathogen growth by creating physiologically restrictive environment through the release of antimicrobial factors such as lactic and acetic acid, hydrogen peroxide, defensins and bacteriocins as discussed previously; (d) probiotic induced enhancement of mucosal barrier function; and (e) induction of T-cell apoptosis in the lamina propria. However, it remains to be established that which of the above mentioned potential mechanism relate to the treatment of IBD and how the pathophysiology of clinical disease is altered through consumption of probiotics.

Beneficial effects of probiotic bacteria may be broadly classified into two main categories: (I) those effects arising due to activity in the large intestine related to colonization and inhibition of pathogen growth and (II) those effects related to enhancement of the host immune response and barrier function through interactions with epithelial and immune cells within both the small and large intestine. Most of the probiotic bacteria confer the effects mentioned in the first category. However, the effects in the second category are species specific and seem to be very essential in treating IBD. Cytokines play an important role in the growth, activation and differentiation of immune cells. Cytokines also regulate the process of inflammation and immune tolerance. Imbalance between Treg/Th1, Th2, and Th17 affect the deletion or apoptosis of over- and autoreactive cells. The lack of Treg cells or the excessive production of effector T cells participates in the development and exacerbation of IBD. $CD4^+$ lymphocytes with Thl/Th17 are found to be dominant in mucosal sections of Crohn's disease where as $CD4^+$ lymphocytes with Th2 phenotype in UC. Therefore, Thl and Th17 become targets to treat Crohn's disease and Th2 in preventing UC. As discussed in the earlier chapters, intestinal bacteria can modulate the immune system at both systemic and mucosal level. In mice models, *L. plantarum* has proven to decrease IL- 12, IFN-γ, and IgG2a at the mucosal level. Further, consumption of conventional yoghurt produced using a pool of potentially probiotic strains of *L. bulgaricus* and *S. thermophilus* exerted anti-inflammatory effect by diminishing IL-17 and IL-12 while increasing IL-10 levels in the intestinal tissues. In addition, LAB downregulate pro inflammatory cytokines by inhibiting the epithelial cell NF-κB pathway. These *in vivo* studies suggest that immunomodulatory properties of probiotics through the repression of pro inflammatory cytokines could be one of the beneficial aspects that enable mainly LAB and other probiotics to prevent and treat IBD.

Further, LAB that produce antioxidant enzymes can also be used in the treatment of IBD. IBD is frequently associated with oxidative stress and epithelial damages due to recurrent and abnormal inflammation. Several studies explained the correlation between increase in the reactive oxygen species (ROS) production and disease activity in inflamed biopsies of IBD patients. Therefore probiotics capable of degrading ROS by expressing antioxidant enzymes or at least interrupting their formation could be used in alleviating the symptoms of IBD. ROS are byproducts of oxygen metabolism and at low concentrations they are essential since they involve in cell signaling and regulatory pathways. But when they produce at higher concentrations, they act to eliminate infectious agents by causing severe damages to cell structures and macromolecular constituents such as DNA, RNA, proteins and lipids. To combat this effect, the normal intestinal mucosa of healthy individuals has well furnished with ROS neutralizing antioxidant enzymes such as catalase, glutathione peroxidase, glutathione reductase, glutathione-s-transferase, and superoxide dismutase. Whereas in case of IBD patients, levels of these enzymes are frequently depleted. Therefore, probiotic LAB strains expressing high levels of antioxidant enzymes could prevent oxidative epithelial damages by increasing their activities in specific locations of the gastrointestinal tract of IBD patients.

Table 1. Lists some examples of probiotics with proven anti-inflammatory properties classified by their mechanisms of action.

Mechanism	Strain	Demonstrated effect
Immune	*B. longum*	Improvement of clinical appearance of chronic inflammation in patients, decreases in TNF-α and IL-1α.
	BIFICO (3 bifidobacteria species)	Prevention of flare-ups of chronic ulcerative colitis,inactivation of NF-κB, decreased expressions of TNF-$\alpha$and IL-1β and elevated expression of IL-10.
	L. salivarius ssp. *salivarius* CECT5713	Recovery of inflamed tissue in TNBS model of ratcolitis, increase in TNF-α and iNOS (inducible NOsynthase) expression.
	L. fermentum, *L. reuteri*	Improvement of histology in a TNBS model of ratcolitis, decreased levels of TNF-α and i-NOS expression.
	L. casei Shirota	Improvement in murine chronic inflammatory boweldisease, downregulation of proinflammatory cytokinessuch as IL-6 and IFN-γ.
	L. casei DN-114 001	Reduction in numbers of activated T lymphocytes inthe lamina propria of Crohn's disease mucosa, decreaseof IL-6 and TNF-α.
	L. plantarum 299v	Decreased IL-12, IFN-γ and IG2a at the mucosal levelof specific pathogen free IL-10 KO mice. Decreasedmesenteric lymph node IL-12 and IFN-γ production aswell as histologic colitis scores in the pretreatment ofGF mice that were exposed to normal flora.
	L. rhamnosus GG	Alleviating intestinal inflammation, decrease TNF-α. Specific inhibition of macrophages TNF-α productionby a contact independent mechanism.
	L. salivarus UCC118	Reduced production of proinflammatory cytokines inIL-10 KO mice injected subcutaneously with theprobiotic strain.

Altering the microbiota	*L. salivarus*UCC118	Reduction of *Clostridium perfringens*, coliforms and enterococcus Levels in IL-10 KO mice. Production of a peptide that inhibits a broad range of pathogens such as *Bacillus, Staphylococcus, Enterococcus, Listeria* and *Salmonella* species.
	L reuteri	Decreased concentration of colonic *Lactobacillus* species and increased concentration of mucosaladherent bacteria associated with colitis attenuation.
	VSL#3	Delayed relapse into pouchitis after surgical resection inhuman patients. Decreased luminal pH in patients with UC that antagonizes pathogenic bacteria. Reduced counts of bifidobacteria and lactobacilli in Pouchitis. Delayed relapse into pouchitis after surgical resection.
Unknown	*L. rhamnosus GG, L. rhamnosus Lc705, P. freudenreichii ssp. shermanii JS and B. breve Bb99*	Alleviating irritable bowel syndrome symptoms
	L. plantarum	Improvement of the disease activity index in aninduced rat colitis model
	*L. rhamnosus*GG	Improvement in the clinical status in children withmildly to moderately active stable Crohn's disease
	*L. casei*Shirota	Improvement in the clinical condition of murin DSSmodel of ulcerative colitis

Lactose intolerance

Lactose, the predominant carbohydrate in milk, is a disaccharide consisting of galactose bound to glucose and serve as the major source of energy for new born. Intestinal absorption of lactose requires hydrolysis to its component monosaccharides by the brush-border enzyme lactase. Thus hydrolysed glucose and galactose are taken up by the intestinal cells and transported into the bloods tream. A considerable part of glucose and most galactose is cleared by the liver after the first pass. Lactose which is not hydrolyzed in the small intestine passes into the colon where it is fermented by colonic bacterial enzyme β-galactosidase. Where, glucose and galactose are subsequently converted into lactate as well as into the short chain fatty acids (SCFA) acetate, propionate and butyrate. Lactose and its metabolites are osmotic active products causing an osmotic pressure which further increases the colonic transit time and leads to the development of clinical symptoms as diarrhea.

Lactase enzyme activity is affected by various factors like aging, genetic background, integrity of the small intestinal membrane and the small intestinal transit time upon which the turnover of lactose by the enzyme is dependent. The decreased lactase activity results in gastrointestinal symptoms including diarrhea, bloating, flatulence, and abdominal discomfort. The term *lactose intolerance* indicates lactose malabsorption causing these gastrointestinal symptoms.

Pre- and probiotics can be used to alleviate the symptoms of lactose intolerance. The hydrolytic ability of the probiotics strains can be used to reduce the amount of lactose in the milk and milk products with lactose as occurs in yogurt. Probiotics can also be used to increase the overall hydrolytic capacity in the intestine. It has been showed that consumption of yogurt containing *Lactobacillus bulgaricus*and *Streptococcus thermophiles* alleviate the lactose intolerance through their enzyme activity when the product reaches the intestinal tract. Allevation of the symptoms of lactose intolterance can also be achieved by manipulating the colonic metabolism. However, the precise mechanism how colonic metabolism influences lactose intolerance symptoms is not yet known.

Celiac disease

Celiac Disease (CD) is a form of chronic enteropathy affecting the small intestine in genetically predisposed individuals and is precipitated by the ingestion of gluten containing foods. It is also referred as gluten sensitive enteropathy, celiac sprue and non-tropical sprue. The disease is characterized by severe immune mediated damage to the jejuna mucosa, typically involving chronic diarrhea, abdominal bloating and distention, weight loss, iron deficient anemia, malnutrition and metabolic bone disease, partial to subtotal villous atrophy. Trigger from gluten protein of wheat and related cereal, presence of Human Leukocytic Antigens mainly HLA-DQ2 or HLA DQ-8 molecules and the generation of circulating autoantibodies to tTG (tissue transglutaminase) are essential to precipitate the disease. Gluten a wheat protein mainly constitutes glutenin and gliadin fractions rich in proline and glutamine residues. Since gastric proteases lack post prolyl activity, upon ingestion of gluten, immunogenic gluten peptides capable of resisting the activity of gastrointestinal proteases are generated and reach lamina propria either by epithelial transcytosis or by increase in the epithelial tight junction permeability. Then glutamine residues in the peptides undergo deamidation by tTG expressed just beneath the gut wall epithelium. Deamidated gluten peptides with more negatively charged residues bind to HLA-DQ2 or HLA-DQ8 molecules on antigen presenting cells with great affinity. tTG Deamidated gluten residues presented by HLA-

DQ2 and DQ8 molecules turn on the gliadin reactive CD4 T_H cells. T cells further trigger humoral mediated immune pathway and activates B cells to produce antibodies against gluten peptides, tTG and also produces array of cytokines: interferon-γ, interleukin (IL)–1β, tumor necrosis factor–α, IL-6 and IL-8 at higher levels. These cytokines make the enteric lymphocytes as cytotoxic cells and affects the enteric cells resulting in inflammation.

Celiac disease occurs as the result of interplay between the genetic tendency of certain individuals to develop disease and excessive and early ingestion of gluten. Nutritional therapy is the only treatment for CD, which involves life time elimination of wheat, rye and barley from the diet. Several studies revealed that follow-up of strict gluten free diet ameliorated the pathogenicity of CD. Predisposition can be controlled by breeding less immunogenic varieties of wheat and by biotechnological approaches. Since microbial proteases are known for efficient proteolytic activity they are extensively used to reduce gluten in the food for celiac patients. But major problems associated with the use of these proteases include the *in vivo* efficacy of the microbial or enzyme preparations, efficient delivery against gastric conditions in the stomach, formulation and dosage.

Proteolysis by probiotic isolates has been suggested as a new tool for food processing for celiac persons. Therefore more emphasis must be given to improve gluten hydrolyzing probiotic isolates such that gluten reduced wheat and other related cereal foods, oral enzyme supplementation of glutenases could be the promising for celiac prone individuals.

However, VSL#3 (a mixture of 4 lactobacilli strains *L. plantarum, L. casei, L. acidophilus*, and *L. delbrueckii ssp. bulgaricus*), a probiotic preparation effectively reduced the gluten toxicity during long term wheat flour fermentation. When jejuna biopsies of celiacs were exposed to peptic- tryptic digest from VSL#3 fermented dough, there was no increase in the infiltration of $CD3^+$ intraepithelial lymphocytes and zonulin release reduced when biopsy samples were exposed to VSL#3 treated gliadins. In an *in vivo* study after ingestion of gluten pretreated with ALV003 (preparation of mixture of highly specific glutanases) to CD patients, gluten specific T cell response was reduced compared to placebo treated gluten. These studies gave an idea of using microbial proteases as a potential tool for the treatment of CD either by oral protease supplementation or by development of gluten reduced food stuffs.

Key questions

1. Antibiotic associated Diarrhea is the most common gastrointestinal diseases along with side effect of antibiotic therapy it is mainly associated with.

    a) *Helicobacter pylori* b) *Clostridium difficile*

    c) *Escherichia coli* d) *Salmonella typhi*

2. The probiotic metabolites important in controlling establishment of *H. pylori* infection in humans are.

    a) Bacteriocins and SCFA

    b) Mucins and adhesions

    c) Exopolysaccharides and oligosaccharides

    d) Antibacterial agents and defensins

3. Probiotics in inhibiting traveler's diarrhea.

4. Role of cytokines produced by probiotics in treating IBD.

Further readings

Al Kassaa, I., & AL KASSAA, I. (2017). Antiviral probiotics: a new concept in medical sciences. New insights on antiviral probiotics: From research to applications, 1-46.

Fassio, F., Facioni, M. S., & Guagnini, F. (2018). Lactose maldigestion, malabsorption, and intolerance: a comprehensive review with a focus on current management and future perspectives. Nutrients, 10(11), 1599.

Guandalini, S., & Discepolo, V. (2022). Celiac disease. Textbook of pediatric gastroenterology, hepatology and nutrition: A comprehensive guide to practice, 525-548.

Kopacz, K., & Phadtare, S. (2022, August). Probiotics for the prevention of antibiotic-associated diarrhea. In Healthcare (Vol. 10, No. 8, p. 1450). MDPI.

Kumar, K. S., Sastry, N., Polaki, H., & Mishra, V. (2015). Colon cancer prevention through probiotics: an overview. J Cancer Sci Ther, 7(2), 081-092.

Olin, S. J., & Bartges, J. W. (2015). Urinary tract infections: treatment/comparative therapeutics. Veterinary Clinics: Small Animal Practice, 45(4), 721-746.

7

Genome Sequences: Approaches to Genome Wide Analysis of Gut Bacteria

Diversity of gut microbial community

The human microbiota (the collection of microbes that live on and inside us) consists of about 100 trillion microbial cells that outnumber our "human" cells 10 to 1, and that provide a wide range of metabolic functions that we lack. It has been estimated that they encode 100-fold more unique genes than our own genome. Furthermore, the gut microbes contribute to energy harvest from food, and changes of gut microbiome may be associated with bowel diseases or obesity. To understand and exploit the impact of the gut microbes on human health and well-being it is necessary to decipher the content, diversity and functioning of the microbial gut community. If we consider ourselves as superior organisms encompassing these microbial symbionts, by far the majority of genes in the system are microbial. In this sense, completing the human genome also requires the characterization of the microbiome (the collection of genes in the microbiota).

Methods of counting viable and non-viable bacteria in gut

Before the advent of genome sequencing and genome-wide analysis method, our knowledge of molecular characteristics of diverse gut microbial population was extremely limited. To obtain an estimate of cultivability, microbial ecologists generally compare microscopic counts with total viable counts. The total viable count made on a nonselective agar based medium estimates the number of colony forming units (cfu) per gram of sample. The finding that total viable counts are typically lower than total microscopic counts was thought to be due to the number of dead cells. Indeed, dead bacteria in feces may constitute up to one third of the total bacterial community. However, recent nucleic acid based studies indicated that a majority of bacteria in a variety of ecosystems are different from those described in culture. Therefore, it is reasonable to assume that the differences between microscopic and total viable counts are also due to an inability to culture the majority of the bacteria. Culture dependent methods allow characterizing only the countable part of the microorganisms in the gut sample. Moreover, lack of morphological details in many bacteria and other microorganisms has usually hampered their identification, making the use of cultivation necessary for identification. This has proven difficult for complex microbial environment like gut.

Techniques for enumeration of cultivable and non-cultivable microorganisms

Traditional methods to determine the composition of microbiota require cultivation on suitable selective media, which is laborious, time- consuming and prone to statistical and methodological errors. According to classical microbiology, in order to study a microorganism, it is necessary to isolate it from the original matrix, and this isolation can be performed only on the appropriate medium, from the plates showing distinct colonies i.e., from the plates corresponding to the so-called countable dilutions. Moreover, it has been demonstrated that environmental modification during cultivation could affect the structure of microbial communities, thus preventing a complete view of the ecosystem considered. Therefore, isolation and cultivation in agar medium allow characterizing only those microorganisms capable of growing, multiplying and forming colonies on the selected medium and condition of growth, with the loss of the information relative to the microbial component which presents a vitality condition despite not being able to duplicate in culture medium. Although new microorganisms are continuously isolated, it is estimated that only a small fraction of the existent microorganisms have been grown in pure culture and characterized. Lack of knowledge is most severe particularly for complex multi-species gut microbial communities. Suppose even when all bacteria could ultimately be cultured (which is quite unlikely), progress in the understanding of the ecology of complex microbial communities would require studies on the activity and distribution of microbes directly in minimally disturbed samples. This has led to an extensive development and application of culture-independent approaches to study complex microbial ecosystems.

Genome based insights for identification of microorganisms

Genome based techniques provide insight into both cultivable and non cultivable microorganisms. High-throughput sequencing studies and new software tools are revolutionizing microbial community analyses, yet the variety of experimental and computational methods can be daunting.Technological developments in the past 50 years in genomics, transcriptomics, proteomics, metabolomics and bioinformatics contributed a lot to the genome wide analysis of gut bacteria. The integrated use of these technological approaches can be implemented to elucidate underlying molecular mechanisms of important cellular processes and to discover novel factors involved diseases, the pathogenicity of microbes, resistance to antibiotics, and so on. To date, research on gut microbes has greatly benefited from these high throughput toolkit and approaches. The full genomic microbial DNA sequences available to date (e.g. see http://www.tigr.org/tigr-scripts/CMR2/CMRHomePage.spl) enables a comparative genomics approach for which various novel bioinformatics tools and database systems have been developed. This includes the (partial) construction of metabolic pathways for the organism of interest, superimposing gene regulatory information, the automatic discovery of important regulatory *cis* elements in intergenic regions, the discovery of protein function using a wide variety of tools (e.g. domain, operon, next neighbour and metabolic pathway analyses) and linking this information to the available literature. These efforts have already yielded valuable databases and network information for bacteria such as *Bacillus subtilis*, *Lactococcus lactis*, *Lactobacillus plantarum* and various pathogens. Even more importantly, many efforts have now been started in the use of experimental high-throughput approaches that enable transcriptome, proteome and metabolome determinations. This adds a dynamic level to the description of food and gut bacteria, as these are known to vary gene expression, protein synthesis and metabolite production significantly and rapidly during growth, and even more in response to changing growth conditions (e.g. changes in nutrients, pH, temperature, oxygen etc). However, these studies are challenging because of the scale and complexity of the microbiome and also due to the unexpected variability between individuals.

Currently, there are two main methods for performing human microbiome characterization that do not rely on growing organisms in pure culture: small-subunit ribosomal RNA (rRNA) studies, in which the 16S rRNA gene sequences (for archaea and bacteria) or the 18S rRNA gene sequences (for eukaryotes) are used as stable phylogenetic markers to define which lineages are present in a sample and metagenomic studies, in which community DNA is subject to shotgun sequencing. Metagenomics is the study of genetic material recovered directly from environmentally derived samples without laboratory cultures. Metagenomic studies involves the isolation of DNA samples from the environmental samples, amplifying the collective of 16S rRNA genes with PCR primer sets, subcloning the resulting PCR product and classifying the taxa present according to a database of assigned 16S rRNA sequences. Alternatively, DNA or RNA can be isolated, sub cloned and then sequenced to produce a fragment pool representative of the existing population. These sequences can be then translated into protein fragments and compared with the existing database of published genome sequences to identify community members. Recent studies use “shot gun” Sanger sequencing or massively parallel pyrosequencing to get largely unbiased samples of all genes from all the members of the sampled communities. As the price of DNA sequencing continues to fall, metagenomics now allows microbial ecology to be investigated at a much greater scale and detailed than before. Metagenomic sequencing represents a powerful alternative to rRNA sequencing for analysing complex microbial communities. Small subunit rRNA-based studies are sometimes also considered to be "metagenomic" in that they analyze a heterogeneous sample of community DNA.

Sequence comparisons of nucleic acids isolated from complex microbial ecosystems can be used to provide molecular characterization, while at the same time providing a classification system, which predicts natural evolutionary relationships. Community profiling, or determining the abundance of each kind of microbe, is much cheaper using rRNA because only one gene out of each genome is examined but metagenomic profiles are essential for understanding the functions encoded in those genomes. Community profiling can provide information about identity, distribution and abundance of microorganisms; similarities and/or changes in communities over time or between sites. Techniques that probe gene expression directly such as meta transcriptomics and meta proteomics (analysis of the transcripts or proteins in a community, respectively), although useful in simpler microbial communities such as acid mine drainage are just beginning to be applied to human-associated microbial communities. Metagenomics and r-RNA based techniques contributed to the progress made in characterizing the human microbiome and its role in health and disease in the past few years, particularly with the advent of high-throughput sequencing. 16S rRNA sequence based methods revealed that two bacterial divisions Bacteroidetes and Firmicutes constitute dominate known phylogenetic categories of the distal gut microbiota. Studies also revealed the existence of substantial diversity of gut microbiome between healthy individuals. This diversity is marked among the infants compared to adults where in adulthood the gut microbiome converges to more similar phyla.

Many ongoing Human Microbiome Projects (HMPs) have raised a key question that whether there is a core human microbiome of genes or species that we all share. If so, the strategy for understanding the microbiome is clear: Identify the organisms that comprise the core using 16S rRNA analysis, sequence their genomes, and use these genomes as scaffolds for metagenomic, metatranscriptomic, and metaproteomic studies that provide information about small fragments of genes, transcripts, or proteins, respectively, but that require assembly against known sequences. However, if there is a minimal core or no core at all, alternative strategies will need to be developed because new genes and species will continue to be found in each new person examined.

Sequencing of Small Sub Unit (SSU) rRNA genes has become a standard procedure in the identification of isolates, and it is now impossible to adequately describe microbial communities without SSU rRNA sequence data. Hence, construction of SSU rRNA libraries is required to invent bacteria and archaea present in a given environment. Currently, > 79,000 16S rRNAsequences are available in DNA databases, which is far greater than for any other gene (http:// rdp.cme.msu.edu/html). rRNA sequences can be obtained either directly from SSU rRNA or from their encoding genes (SSU rDNA) by reverse transcription (RT)- or regular PCR. In practice, SSU rDNA sequences are determined by creating rDNA clone libraries rather than cDNA libraries from rRNA. After construction of library, the sequences of the cloned amplicons are determined and compared to sequences deposited in SSU rDNA databases (http: //www. ncbi. nlm.nih.gov/BLAST/ and http://rdp.cme.msu.edu/html), and followed by phylogenetic analysis. Though sequencing of cloned SSU rDNA amplicons gives significant knowledge regarding the identity of uncultured bacteria, the data are not quantitative and PCR based cloning steps are with biases. Nevertheless, the SSU cloned libraries to date contributed a lot to better understand the diversity of the gut microbiota.

Types of genome sequencing methods and its applications

Community genomics, the use of genomes sequenced from natural microbial communities to understand the structure and metabolism of the community, has been successful in environments with varying levels of diversity. Recently, this approach has been implemented to the human microbiome. Increased sequencing depth allowed for genomes to be assembled for both high abundance and low-abundance members of the microbial community found in the gut. Strain-level analysis of microbial communities contrasts strongly with 16S rRNA-based fingerprinting methods that characterize communities with phylum to genus-level resolution. This is primarily due to the added benefit of being able to directly determine the metabolic potential of strains in a particular community, and to identify metabolic variation between strains that may have highly similar or even identical 16S rRNA gene sequences.

Microbial community profiling using 16S rRNA currently undergoing revamp as high throughput techniques such as barcoded pyrosequencing allow to gain deep views into hundreds of microbial communities simultaneously. Pyrosequencing is a method of DNA sequencing based on the "sequencing by synthesis" principle and differs from Sanger sequencing method. The desired DNA sequence is able to be determined by light emitted upon incorporation of the next complementary nucleotide by the fact that only one out of four of the possible A/T/G/C nucleotides are added and available at a time so that only one letter can be incorporated on the single stranded template (sequence to be determined). Incorporation of each nucleotide by DNA polymerase results in the release of pyrophosphate, which initiates a series of downstream reactions that ultimately produce light by the firefly enzyme luciferase. The intensity of light determines if there is more than one of these letters in a row. The previous nucleotide letter (one out of four possible dNTP) is degraded before the next nucleotide letter is added for synthesis: allowing for the possible revealing of the next nucleotide (s) *via* the resulting intensity of light. This process is repeated with each of the four letters until the DNA sequence of the single stranded template is determined.

In pyrosequencing, it is possible to run millions of molecules in a single run, but it is still pretty expensive. When the bacterial composition of complex ecosystem such as gut has to study by looking at their 16S rRNA genes, it would be very expensive to use a single pyrosequencing run per sample. In this context, barcoded pyrosequencing has been developed. It is one of the ways to combine the 16S rRNA gene amplication of different samples and to run them in a single pyrosequencing run. In this method, at first different samples to be sequenced are amplified with a different primer set. One part of the primer is the universal primer and is same for all the samples, while; other part (usually a 3 to 5 letter code) varies per sample and generates the variation essential to distinguish between samples. These studies are made possible by the remarkable observation that a small fragment of the 16S rRNA gene is sufficient as a proxy for the full-length sequence for many community analyses, including those based on a phylogenetic tree. Moreover, labor and expense make SSU rDNA sequences unsuitable for monitoring communities in a culture independent fashion. However, fingerprinting techniques (techniques that provide limited information about the microbial community) are still orders of magnitude cheaper and faster to perform. In contrast to cloning and sequencing, fingerprinting techniques enable the rapid analysis of large sample numbers including replicates. They appear to be ideal for monitoring community shifts and comparing communities between different gastrointestinal regions.

Some fingerprinting techniques include Denaturing Gradient Gel Electrophoresis (DGGE), Temperature Gradient Gel Electrophoresis (TGGE), Temporal Temperature Gradient Gel Electrophoresis (TTGGE), Single Strand Conformation polymorphism (SSCP) and Terminal- Restriction Fragment Length Polymorphism (T-RFLP).These methods rely on the amplification of a specific gene, typically 16S r RNA but not always and then separating different variants of the gene in the community sample by electrophoresis. The principle of these finger printing techniques varies and moreover they are PCR-based and generate profiles representing the sequence diversity within the selected ecosystem. TGGE and DGGE are forms of electrophoresis which use either a temperature or chemical gradient to denature the sample as it moves across an acrylamide gel. In case of DGGE, nucleic acids migrate in a chemical gradient according to their GC content. Whereas TGGE relies on temperature dependent changes in the structure nucleic acids. DGGE was the original technique, and TGGE a refinement of it. After introduction of TTGGE and DGGE, the face of microbial ecological research has been changed. These two techniques are regularly used in the exploration of microbial communities of different environments. TTGGE exploits the principle on which DGGE is based without requiring a chemical denaturing gradient. DGGE, TGGE, and TTGGE analyses of SSU rDNA is now common and have been used successfully to characterize and monitor GI bacterial communities in humans. Whereas SSCP relies on the on the secondary structure of single stranded DNA and is defined as conformational difference of single stranded nucleotide sequences of identical length as induced by differences in the sequences under certain experimental conditions. This property allows sequences to be distinguished by means of gel electrophoresis, which separates fragments according to their different conformations. Although, less frequently used T-RFLP has also proven to be a useful fingerprinting technique to monitor the GI microbiota.

T-RFLP is a molecular biology technique for profiling of microbial communities based on the position of a restriction site closest to a labeled end of an amplified gene. The method is based on digesting a mixture of PCR amplified variants of a single gene using one or more restriction enzymes and detecting the size of each of the individual resulting terminal fragments using a DNA sequencer. The result is graph image where the X axis represents the sizes of the fragment and the Y axis represents their fluorescence intensity.

Fluorescence *in situ* hybridization (FISH) is another molecular technique that uses fluorescent nucleotide probes to detect and localize specific DNA- or RNA sequences. This technique is considered as a powerful tool for phylogenetic, ecological, diagnostic and environmental studies in microbiology. In contrast to PCR-based methods, where the spatial structure of the sample is disrupted in DNA/RNA extraction, FISH can be applied to resolve the spatial distribution of microorganisms in a biofilm. When fluorescent probes specific for a functional group of bacteria such as ammonia oxidizers are used, the comparison of probe-based enumeration with nitrification rates also provides a basis on which to measure the specific in situ activity per single cell. Application of FISH to intestinal biopsies showed the spatial distribution of bacteria in the gut lumen and the mucus layer. Through the use of species specific probes in FISH, it is possible to identify different microorganisms in complex microbial communities, thus providing a solid support to the understanding of inter species interaction. Probes used for FISH target small regions of the RNA. This may restrict their specificity and make taxonomic identification difficult in complex microbial environment such as gastro intestinal tract, which harbors a large fraction of previously uncultured and undescribed species with strictly anaerobic and symbiotic growth requirements. Moreover, so far only few fluorophores have been identified for FISH. This limits the simultaneous identification of many numbers of bacteria in a single FISH experiment. In such cases, to make the FISH more efficient FISH-MAR is used. It is the combination of FISH with micro autoradiography in order to fuse phylogenetic information with radioactive labeling. FISH-MAR can be efficiently used to investigate the ecophysiology of defined organism of interest in detail, for which a FISH probe is available. In addition, FISH-MAR is very sensitive technique that requires short incubation time and reduces the chance of false positive results due to metabolic cross-feeding.

BOX-PCR is another DNA based typing method known for its simplicity, efficiency and low cost. This is a specific version of repetitive extragenic palindromic-PCR (rep-PCR) that uses the BOX-A1R primer. BOX-PCR is fingerprinting analysis based on the BOX-dispersed repeat motif, firstly identified in *Streptococcus pneumonia*, but common in a number of bacterial species. Since the BOX repetitive sequences are interspersed throughout the genome, BOX-PCR is a method potentially capable of simultaneously surveying many DNA regions scattered in the bacterial genome. It has been shown to have similar or even better strain differentiation power, as well as to be easier to perform than, ribosomal intergenic spacer analysis (RISA), RFLP, amplified fragment length polymorphism (AFLP), random amplified polymorphic DNA (RAPD) and other techniques. BOX-PCR is quicker, cheaper and in many cases it was found more discriminatory than pulse field gel electrophoresis (PFGE). BOX-PCR patterns are not affected by the culture age of the strain to be analyzed and output can be easily analyzed by computer assisted methods.

The interpretation of finger printing techniques available for the genome wide analysis of the gut microbiota is very difficult due to the vast microbial diversity. A number of methods have been used for the quantitative analysis of finger print patterns and comparision of communities. Recently, statistical models are fetching the interest of microbial ecologists, but such methods are still under development. Compared to statistical methods available for plants, animals and human epidemiological and biomedicine studies, the methods for microbial community studies are yet in the preliminary stage of development. Almost all methods suffer from inevitable limitations including bias resulting from samples run on different gels or machines, PCR *bias* or low quality amplification products; differences in detection limits, which affect the interpretation of false and true bands or peaks and also lack of distinction between closely adjacent bands or peaks. Looking into these lacuna, it is quite evident that different investigators end up with slightly different fingerprint patterns interpretation. Nonetheless, computational tools such as principal components analysis (PCA), functional principal component analysis (FPCA), clustering of correlations, self organizing maps (SOM) or partial least squares regression (PLS-R) have been found to be useful in the interpretation of fingerprinting datasets.

Table 1. A summary of current techniques used to study complex microbial ecosystems

Methods	Uses	Limitations
Cultivation	Isolation; ideal	Slow and laborious, not applicable for non cultivable microorganisms,.
16S rDNA sequencing	Phylogenetic Identification	Laborious; subject to PCR biases
DGGE/TGGE/ TTGGE	Monitoring of community/ population shifts; rapid comparative Analysis	Subject to PCR biases; Semi-quantitative; identification requires clone library
BOX-PCR	Rapid, simpler, efficient	Poor band resolution on agarose gel (amplified products subjected to separation by gel electrophoresis).
T-RFLP	Monitoring of community shifts; rapid comparative analysis; very sensitive; potential for high throughput	Subject to PCR biases; semi-quantitative; identification requires clone library
FISH	Detection; enumeration; comparative analysis possible with automation	Requires sequence information; laborious at species Level
Dot-blot hybridization	Detection; estimates relative abundance	Requires sequence information; laborious at species level
Quantitative PCR	Detection; estimates relative abundance	Laborious
Diversity microarrays	Detection; estimates relative abundance	In early stages of development; expensive
Non-16S rRNA Profiling	Monitoring of community shifts; rapid comparative analysis	Identification requires additional 16S rRNA-based approaches

Key questions

1. Use of experimental high-throughput approaches enables determination of
   a) Transcriptome, proteome and metabolome.
   b) Sequence of nucleotides in a DNS segment.
   c) Sequence of DNA and RNA to provide in-depth information about genomic variations
   d) Sequencing and reassembling large genomes like human genome.
2. PCA stands for
   a) Patient controlled analgesia
   b) Principle components analysis
   c) Plate count agar
   d) Proto catechuic acid
3. What is pyrosequencing?
4. Methods for performing human genome characterization.

Further reading

Ayrapetyan, M., & Oliver, J. D. (2016). The viable but non-culturable state and its relevance in food safety. Current Opinion in Food Science, 8, 127-133.

Brown, T. A. (2023). Genomes 5. CRC Press.

Chen, P. E., & Shapiro, B. J. (2015). The advent of genome-wide association studies for bacteria. Current opinion in microbiology, 25, 17-24.

Kwong, W. K., & Moran, N. A. (2015). Evolution of host specialization in gut microbes: the bee gut as a model. Gut microbes, 6(3), 214-220.

Mahon, C. R., & Lehman, D. C. (2022). Textbook of Diagnostic Microbiology-E-Book: Textbook of Diagnostic Microbiology-E-Book. Elsevier Health Sciences.

Pevsner, J. (2015). Bioinformatics and functional genomics. John Wiley & Sons. DOI:10.1002/9780470451496, 1-951.

Tannock, G. W. (2017). Understanding the gut microbiota. John Wiley & Sons.

Wang, T., Birsoy, K., Hughes, N. W., Krupczak, K. M., Post, Y., Wei, J. J., ... & Sabatini, D. M. (2015). Identification and characterization of essential genes in the human genome. Science, 350(6264), 1096-1101.

8

Probiotics Enzymes, Industrial Applications

Introduction

Let food be thy medicine and medicine be thy food- the age-old quote by Hippocrates, is certainly the tenet of today. Microbial cultures have been used for millenniums in food and alcoholic fermentations and in the past century have undergone scientific scrutiny for their ability to prevent and cure a variety of diseases. Later it became popular in improving human and farm animal nutritive values and recently in aquaculture also. The growing interest in the understanding the role of food in human health has moved from its primary role as a source of energy to the subtle action of biologically active food components on human health. Henceforth functional foods or nutraceuticals are of great demand in present time. The growing scientific evidence suggests that the food supplements containing beneficial bacteria can provide an array of benefits to the host. Now a day's probiotic fermented foods are in great demand due to the awareness of beneficial aspects of probiotics among people. Several industries focused on the improvement of conventional fermented foods scientifically in order to fetch good market for their products. Particularly, animal feed companies and researchers have been looking for alternative products and strategies that can help to maintain animal gut health in order to prevent or reduce the prevalence of pathogens in the food chain. After understanding the mechanism and beneficial aspects of probiotics in detail at molecular level in earlier chapters, now it is relevant to discuss the applied part of probiotics. In this chapter, commercial importance of probiotics and probiotics derived products at industrial level are emphasized.

Probiotics synthesized enzymes and its applications

Probiotics are hub of enzymes that improves not only gut but overall health status of an individual. They are known to produce hydrolytic enzymes such as amylase, lipase, array of proteases (capable of cleaving peptides which are resistant to gastric enzymes), cellulose, phytase, lactase and many more that improves digestion and nutrient absorption. Enzymatic equipment enables them to break down a wide variety of food constituents that cannot be metabolized by the host such as galactooligosaccharides, inulin, resistant starches, and antinutritional factors such as tannins or phytates responsible for the chelation of minerals including iron, zinc, magnesium and calcium. Pretreatment of food with these enzymes producing probiotics enhances nutritive quality. In addition, they also have antioxidative properties. Probiotic fermented foods such as yoghurt, acidophilus milk, soy milk, kimchi, kefir, fermented milk beverages, nigerian foods: ogi, ogibaba, wara, kunnu, ugba etc., possess anti-oxidative property. The anti-oxidative property may be due to inhibition of ascorbate auto-oxidation, the scavenging effect of superoxide anion radicals, hydrogen peroxide and also the presence of glutathione system (glutathione is a major cellular non-enzymatic antioxidant. It eliminates lipid- and hydroperoxides, hydroxyl radical and peroxy nitrite mainly *via* cooperation with Se-dependent glutathione peroxidase).

On the other hand, ban on use of antibiotics in animal feeds compelled the industries and farmers as well and increased the demand for probiotics in these industries as an alternative feed additive. The massive use of antimicrobials for disease control and growth promotion in animals increases the selective pressure exerted on the microbial world and encourages the natural emergence of bacterial resistance (World Health Organization antimicrobial resistance fact sheet 194, http://www.who.int/inf-fs/en/fact194.html). Not only can resistant bacteria proliferate after an antibiotic has killed off the other bacteria, but also they can transfer their resistance genes to other bacteria that have never been exposed to the antibiotic. The subtherapeutic (prophylactic) use of antibiotics related to those used in human medicine or the use of any antimicrobial agent known to select for cross-resistance to antimicrobials used in human medicine could pose a particularly significant hazard to human health. Livestock are major reservoir of antibiotic resistant bacteria. Through animal meat they enter human and spread in the society. Contamination of soil and water due to the use of manure from such livestock in agricultural fields also responsible for the entry of antibiotic resistant bacteria to human through crops (Figure 1). With this background, the increasing interest in a healthy diet is stimulating innovative development of novel scientific products in the food industry. The valuable properties of probiotics made them excellent in fulfilling the requirement of functional foods with varying properties.

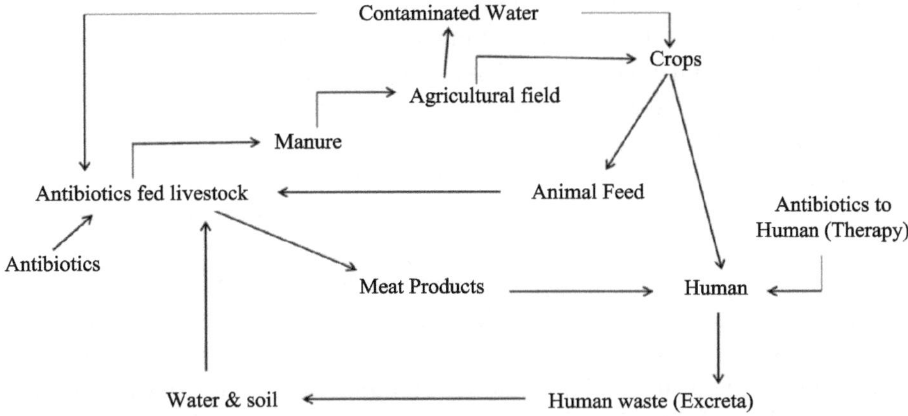

Fig. 1. Possible ways for the spread of antibiotic resistant bacteria & drug resistance genes

Performance Probiotics family of products has been developed using the latest technological advances is microbiology, animal nutrition, immunology and product compounding information. In 1989 FDA adopted the term direct fed microbial (DFM) to refer to a source of live or viable naturally occurring microorganisms derived from bacteria, fungi, and yeast. Over the next few decades vast improvements have been made to demonstrate the positive effects of using organisms and other feed additives in diets to improve animal and human health and positively influence production parameters. DFM bacteria and yeast have been selected from a pool of naturally occurring bacteria and yeast for their ability to positively influence documentable production enhancements. These specifically selected bacteria are incorporated in performance probiotics products for their abilities to aid in digestion, ruminant health and productivity. DFMs are incorporated to the performance probiotics family of products for their ability in improving health status of the host.

Probiotics in aquaculture

Aquaculture of finfish, crustaceans, mollusks, and algal plants is one of the fastest-growing food-producing sectors. Disease outbreaks are being increasingly recognized as a significant constraint on aquaculture production and trade, affecting the economic development of the sector in many countries. Limited success with the use of disinfectants and antimicrobial drugs to prevent or cure diseases of aquatic disease and also the ban on use of the antibiotics in aquaculture due to rejection of export consignments of marine products has raised the use of probiotics in aquaculture as feed additives alternative to antibiotics. Probiotics are extensively used in aquaculture recently to improve the health of shrimps and fishes reducing the use of antibiotics and also as pond cleaners. Probiotic bacteria found to be useful both as feed additive to improve the nutritive status and also as biological control agents of aquatic diseases. Probiotic bacteria directly uptake or decompose the organic matter and/or toxic material and improve the quality of water. The enzymes from probiotics *viz.,* amylase, protease, lipase, xylanase and cellulase in high concentrations play a key role in degrading wastes. Moreover probiotic bacteria have wide range of tolerance to salinity, pH and temperature which usually exists in aquaculture ponds and helps the pond animals to fight against pathogens by their mechanism of action. On contrary, the use of antibiotics disturbs the eubiosis of gut flora eliminating most of the beneficial flora. Upon ending the antibiotic treatment, pathogens began to reestablish themselves. Further invasion of these overgrown pathogens results in inflammatory, immunological, neurological and endocrinological problems. Since probiotics have the ability to fight bacterial and viral infections resisting stress in the ponds, their use replaced antibiotics in aquaculture to a greater extent.

Among several beneficial aspects of probiotics in aquaculture, the growth enhancement property is of high significance. This benefit is thought to occur *via* the colonization of gut by beneficial microflora and their composition. It is postulated that, probiotics increase appetite, improve digestibility and provides source for vitamins, fatty acids, essential amino acids and other micronutrients, the enzymes of probiotics enhance absorbance efficiency of feeds, thus improving growth rate.

Probiotics are usually applied in the feed or added to the culture tank as feed and biological control agents. Most probiotics proposed as biological control agents in aquaculture belong to the lactic acid bacteria (*Lactobacillus, Carnobacterium*, etc.), to the genus *Vibrio* (*Vibrio alginolyticus*, etc.), to the genus *Bacillus*, or to the genus *Pseudomonas*, although other genera or species have also been proved efficient. In commercial hatcheries, eggs and sometimes juveniles of shrimp, mollusk and fish are bathed in bacterial suspension to gain a protective coat of probiotics on to their surface. A relatively dense, nonpathogenic, and diverse adherent microbiota present on their surface would probably be an effective barrier against colony formation by pathogens. In case of fish, larvae may ingest substantial amounts of bacteria by grazing on suspended particles and egg debris. It is therefore obvious that the egg microbiota will affect the primary colonization of the fish larvae. During initial feeding, it is possible to induce an artificial dominance of a certain group of bacteria in the fish-associated microbiota by adding a strain to the rearing water or to the culture medium of the live food. In case of crustaceans, beneficial bacteria are added to the tanks inoculated with diatoms and rotifers to improve survival rate and wet weight of shrimps and crabs. Sometimes, it is impossible to separate feeding aquatic organisms from environmental control. Addition of bacteria to the rearing water of filter feeders such as rotifers, bivalve larvae or adults, and crustacean larvae may result in massive uptake of these bacteria, possibly acting as a food source or contributing to the digestion of the food, even if the main goal of the probiotic application was, e.g., suppression of a pathogen in the culture water.

Maintenance of water quality of the culture pond is crucial aspect in aquaculture. Addition of heterotrophic photosynthetic bacteria may catalyst some important chemical reactions such as nitrogen fixation, oxidation, nitrification, denitrification and sulphurication. In pond water, it also aids in decomposition of various organic wastes *viz.,* remain of food materials, planktons to inorganic phosphate, nitrate and CO_2. These inorganic products again cycle back and aid in the improvement of the nutritive quality and promote microalgae in the pond. Predominance of microalgae is beneficial since it serve as surface for the beneficial bacteria to grow and also as feed for animals. Gram positive bacteria especially *Bacillus* spp.used in ponds are found more efficient in converting organic matter back to CO_2 than Gram negative bacteria, which would favor the conversion of greater percentage of organic carbon to bacterial biomass or slime. The probiotics used in ponds generally have a blend of denitrifying bacteria that diminishes the nitrogen-primary food for algae, making it difficult to form algal blooms at reduced concentration of nitrogen, helps to maintain the number of algae in the pond limiting its overgrowth and contributes to the maintenance of balance between zooplanktons and phytoplanktons. However, sometimes probiotics are added to aquaculture ponds to increase the growth of live foods such as microalgae and rotifers. Live foods are important for marine larvae at early stage of development which usually do not feed on artificial diets. The start up of biofilters by transferring medium from an existing filter is common in aquaculture. It may take some time to get activate. Seeding of such biological filters with nitrifying bacteria effectively reduces the activation period of new biofilters. Although many bacterial cultures containing nitrifying bacteria to control the ammonia level in culture water are available commercially, it is rarely practiced. In addition, the balance between phytoplankton, zooplankton and beneficial bacteria during culture period play a crucial role in the maintenance of pond health.

Several studies reported that many bacterial strains used in pond culture systems have shown algicidal effect on many species of microalgae, particularly of red tide plankton. This antagonistic property of bacteria towards algae would be undesirable in case of larval rearing where algae are used as feed. However, algicidal effect of probiotics may be beneficial in certain cases where there is an undesired algal development in ponds and tanks. Hence it is very important to consider the possible interaction of probiotic bacteria with other zooplanktons and phytoplanktons of the culture system while selecting the probiotic isolates for aquaculture. The selection of isolates for the blend of probiotic species to use in aquaculture varies with the kind of culture method and also the environmental factors. One more process that constitutes the backbone for production yield in aquaculture is reproduction. It is regulated by several factors such as species cultivated, nutrition and environment. Nutrition is closely intermingled with the timed reproductive consequences, from gametes through puberty to adults in both sexes. Recent researches focused on the possible role of probiotic in reproductive process and new progeny with special emphasis to the marine species. Probiotic bacteria used as dietary additives seem to offer an attractive choice inducing overall health benefits to the host organism. Thus the concept of using probiotics or direct fed microbial in animal feed particularly poultry and aquaculture is slowly becoming popular.

Probiotics in poultry

The extensive use of antibiotics in poultry with the purpose of promoting growth rate, increasing feed conversion efficiency to improve meat and egg production and for the prevention of intestinal infections have led to an imbalance of the beneficial intestinal flora, drug residues in the body of birds and the appearance of resistant bacteria. With increasing concerns about antibiotic resistance, there is increasing interest in finding alternatives to antibiotics for poultry production. Natural feed additives, such as live "probiotics" have potential to reduce enteric disease in poultry and subsequent contamination of poultry products. The mode of action of probiotics in poultry includes maintaining normal intestinal microbiota by competitive exclusion and antagonism, altering metabolism by increasing digestive enzyme activity and decreasing ammonia production, improving feed intake and digestion, neutralizing enterotoxins and stimulating the immune system. Addition of probiotics has shown beneficial effects on growth performance of poultry.

Probiotic species belonging to *Lactobacillus*, *Streptococcus*, *Bacillus*, *Bifidobacterium*, *Enterococcus*, *Aspergillus*, *Candida* and *Saccharomyces* have a beneficial effect on broiler performance. These probiotics have beneficial effect on broiler performance, improving sensory characteristics of dressed broiler meat and promoting microbiological meat quality of broilers. Moreover, in broilers supplementation of *Bacillus cereus* and *Saccharomyces boulardii* improved feed conversion rate by 12to 13%. Probiotics significantly improve average live weight and daily weight of the birds and also numerically reduce the mortality rate. In addition, probiotics significantly increase the digestion due to their high proteolytic and amylolytic enzyme activities which aids in macronutrient metabolism. Probiotics in broiler feed also improved the meat quality both at prefreezing and postfreezing storage. Scores for commercially important sensory attributes such as appearance of meat balls, texture, juiciness and overall acceptance are significantly higher for probiotic fed birds. Administration of probiotics is much essential for commercially reared chicken than the wild bird. Because in the wild, the chicken receive complete gut microflora from its mother's feces and would consequently be protected against infection. Whereas in case of commercially reared chicken, the chicks are hatched in incubators which are maintained clean and do not harborany bacteria that are commonly found in the chicken gut. Therefore, an immediate use of probiotics supplementation at birth is more important and useful in avian species than in other animals. The chicken is an ideal example of young animal which is impoverished of its mother contact or adults of its variety immediate after birth, preventing rapid vertical transfer of beneficial microorganisms and results in the disturbed gut microecology. So the supplementation of microbial preparations to restore gut flora at young stage is essential to rear healthy chickens.

Probiotics for farm animals

It is only recently that there is a renewed interest in the use of probiotics for humans and animals and for the understanding of their mode of action. The renewed interest in probiotics has emerged from a general public andscientific concern about the widespread use of antibiotics and the possibility for transfer of antibiotic resistance to human pathogenic bacteria. For this reason the European Union banned the use of antibiotics for non-therapeutic purposes in January 2006. The ban of antibiotics as growth promoters (AGPs) has been a challenge for animal nutrition increasing the need to find alternative methods to control and prevent pathogenic bacterial colonization. The modulation of the gut microbiota with new feed additives, such as probiotics and prebiotics, towards host-protecting functions to support animal health, is a topical issue in animal breeding and creates fascinating possibilities. Man and animals are born with a sterile digestive tract, but very soon after birth a wide diversity of microorganisms begins to colonise the digestive tract. In case of monogastric animals (pigs and poultry), a rich microbiota is present in large intestine whereas in polygastric animals (ruminants), it is in foregut. Digestive microbiota involves in both digestion and modulation of local immune system as well, offering the possibility to exert a positive and completely natural effect on animal health, well being and performance of the animal. Therefore the main target of probiotics is digestive microbiota and its functions as its stability are essential for the health and the nutrition of the host. Mechanism of probiotics in ruminants slightly varies from that of humans; it is mainly due to the difference in the digestive system. The digestive system of ruminants comprises several different prestomachal compartments: reticulo-rumen and the omasum. The abomasum is the "real" stomach of the animal. In ruminants, the foregut does not exhibit any enzymatic activity from the host animal; in fact, the main biological activity in the rumen is from microbial origin. In this large compartment the ecological conditions are well adapted to microbial life (relatively stable pH 6 to 6.8, temperature 37°C). The low redox potential (-400mV) is characteristic for a strictly anaerobic biotope. It is well known that fermentation by rumen microflora plays a key role in ruminant nutrition; this distinct feature lends the ruminant animal several advantages in digestive and metabolic processes over non-ruminants. The main role of the rumen microorganisms is related to the degradation of plant polymers (cellulose, hemicellulose, starch, proteins) ingested by the animal. By this process the microbes provide nutrients which are essential for the host nutrition *viz.*, volatile fatty acids, amino acids, vitamins etc. Rumen is the compartment which facilitates the breakdown of ingested plant material to produce energy which is in available form to the host in the form of volatile fatty acids (VFA) such as acetate, propionate and butyrate. Ammonia produced by the protein degradation corresponds to the source of nitrogen for microbial protein synthesis. These bacteria allow using plant derived polymers as feed components which could not be digested by the endogenous enzymes of the body. In monogastric animals such as pigs also microbiota is abundant and highly diversified and is responsible for digestion of dietary components which have not been previously digested in the upper part of the digestive tract. *Lactobacillus* spp, *Bifidobacteria*, *Streptococcus* spp, *Clostridium perfringens* and *Escherichia coli* are the main microorganisms of the pig digestive tract. Probiotics are known to improve the sensory qualities of pork.

In adult ruminants, probiotics are recommended in situations where there is microbial imbalance *ie.*, dysbiosis, as is the case for dairy cows in the transition period when the diet changes from a high forage-based diet to a high concentrate-based diet. Because concentrates are rapidly fermented in the rumen, they give rise to a rapid accumulation of volatile fatty acids, which contribute to decrease ruminal pH. When ruminal pH is below 6.0, the activity of cellulolytic bacteria is seriously decreased and the number of protozoa declines. Among the microbial changes associated with low ruminal pH is an increase in the number of bacteria that are low pH-tolerant such as *Streptococcus bovis* (a lactate producer) and *Megasphaera elsdenii* (a lactate utilizer). As long as lactate-utilizing bacteria metabolize the lactate that is produced the situation remains under control, although the animals are in sub acute acidosis with a rumen pH between 5.2 and 5.6 and experience fluctuations in dry matter intake resulting in decreased milk production. As pH continues to drop, the lactobacilli population will take over, entailing increased lactate concentration further, it contributes to decrease rumen pH below 5.2 and also results in the disappearance of lactate-utilizing bacteria. This condition is called acute or metabolic acidosis and can lead the animal to death if it remained untreated.

In terms of acidosis prevention and enhancement of growth performance, the inclusion of probiotics in beef cattle diets is perhaps the most adopted practice after ionophores. Although not fully understood, the mode of action of probiotic feeding is believed to be to adapt the rumen to the presence of large quantities of lactic acid, whether directly, by feeding lactate-utilizing bacteria, or indirectly by feeding lactate-producing bacteria, which in turn will stimulate the growth of lactate utilizers as previously discussed. Once lactate-utilizing bacterial count increase, the ability to metabolize lactate derived from rapid carbohydrate fermentation also increases, preventing or reducing the risk of acidosis. Minimizing the incidence of acidosis in feedlot cattle, especially during the transition from high-forage to high-grain diets may lead to improved growth performance. Recent research showed that the use of lactate-utilizing bacterium *Megasphaera elsdenii* as a probiotic has yielded interesting results in acidosis prevention and performance enhancement. Drenching *M. elsdenii* intraruminally has been effective in increasing ruminal pH and decreasing lactate concentrations during a rapid transition from high forage to a high concentrate diet.

The probiotics used in animal nutrition (both monogastric and polygastric) can be divided into three maingroups: lactic acid bacteria (LAB), Bacillus spores and yeasts. Sometimes rarely few species of fungi also used in animal feeds. *Aspergillus oryzae, Lactobacillus acidophilus, L. bulgaricus, L. plantarium, Bifidobacterium bifidium, Pediococcus, Enterococcus, Bacillus, Streptococcus lactis* and *Saccharomyces cerevisiae* are the common organisms in probiotic products for farm animals. LAB have been used in the preparation of silage since ancient periods. Some species of LAB form the main intestinal microbiota and are therefore considered absolutely necessary part of the resident microflora in man and other animals. Probiotics belong to the genera Lactobacilli, Pediococci, Bifidobacteria and Enterococci are important LAB in animal nutrition. LAB confer benefits to host by producing inhibitors and antimicrobial compounds such as SCFAs and bacteriocins, exclusion of pathogens by rapid proliferation, formation of defensive barriers by producing polysaccharides, modulating immune system of the host, influencing the physic-chemical conditions and fat metabolism of host intestine, improving nutrient absorption. *Bacillus* can be also used as probiotic in animal feed due to its sporulation property. *Bacillus* spores are highly resistant to gastric environment and can germinate only in the presence of nutrients, water and warm conditions. In addition, they germinate in the digestive tract and grow as vegetative cells but fail to proliferate rapidly to a greater extent. Hence they are considered transient. As exogenous microorganisms, *Bacillus* probiotics have a high potential for stimulating local intestinal immunity. When the spores undergo germination, metabolic activities reach their maximum level; as a result the metabolites disseminate to the surrounding environment and restrict pathogens. Several strains of yeast *Saccharomyces cerevisiae* have also been used as probiotics in animal feeds. Their mode of action includes: modulation of immune system particularly by stimulating IgA response to pathogens; neutralizing bacterial toxins; strengthening mucosal integrity and intestinal cells; binding to pathogens with the help of mannose receptors and eliminating through feces; enhancing assimilation of nutrients.

In addition, yeasts also generate an anaerobic environment in the gut of rumens by scavenging oxygen, facilitating rumen anaerobic cellulolytic microorganisms. Increased cellulolytic activity aids nutrient digestibility for fiber rich diet. Moreover, certain strains of yeast utilize soluble form of sugars more efficiently than lactate producing bacteria and also act as source of peptides, vitamins, organic acids and cofactors which may be required by the lactate utilizing bacteria and thus limits acidosis. Probiotic products can be administered through water or incorporated in the feed. The benefit of probiotics with respect to health status and performance is expected to be highest in young animals such as piglets and in calves since the gut of young animals is sterile and administering probiotics through water at this stage helps to build up beneficial bacteria much faster than the normal course.

Probiotics for human

For human, probiotics are available both as foods (to improve health) and as drugs (to treat a specific medical condition). Use of probiotics as therapeutic agents is discussed in earlier chapters; in this chapter, let's have a brief overview of various probiotic foods, a part of our daily diet. Of all the probiotic-containing foods normally sold, without doubt the most familiar is yogurt. Yogurt is usually made from milk inoculated with *Streptococcus thermophilus* and either *Lactobacillus acidophilus* or *Lactobacillus bulgaricus*. The milk may be from cows, goats, sheep or even water buffalo. Yogurt is available in innumerable forms and flavors, from low fat chocolate yogurt, to drinkable fruit-flavored goat yogurt, to neon-colored yogurt in squeeze tubes geared toward kids. Like yogurt, kefir is another most widely used probiotic fermented beverage available with different flavors. Kefir can be made from cow's, goat's or sheep's milk, as well as from plant "milks" such as soy, rice or even coconut. The chosen milk is inoculated with kefir "grains"- a mixture of bacteria, including

Lactobacillus kefiri and species of the genera *Leuconostoc, Lactococcus* and others, along with lactose- fermenting and non-lactose-fermenting varieties of yeast. Aside from yogurt and kefir, cultured dairy products are also consumed in large areas of the world and these includes: cultured buttermilk- originally a fermented byproduct of butter-making, but now most commonly used as an ingredient in baking; acidophilus milk- milk inoculated with *Lactobacillus acidophilus*, then fermented; sweet acidophilus milk, in which the culture is introduced but no fermentation follows, reducing the higher acidity and fermented flavor of regular acidophilus milk but still allowing the bacteria to stay alive; lebne- spreadable cheese made from yogurt, popular in middle east; Finnish villi-cultured whole milk and kermavilli-cultured cream; lassi, dahi, maziwalala and chach are enjoyed in South Asian countries. Acidophilus milk has several advantages over ordinary milk possessing low fat content and also has longer shelf life. Acidophilus milk and kefir are considered safe for people with lactose intolerance also. Fermented dairy products have antioxidant, anti carcinogenic properties and when consumed with proper medication, sweet acidophilus milk can be very helpful in the treatment of diseases like typhoid, migraine, urological disorders and infections and even pneumonia while yogurt and yogurt derived products proven to reduce abdominal pain and diarrhea.The low fat sweet acidophilus milk and other culture fermented dairy products are rich in vitamins such as vitamin K, vitamin B1 and folic acid. All these vitamins help in the strengthening of bones and muscles in the body, contributing toimmune system. Probiotic foods also help in balancing the gut microflora increasing the number of beneficial bacteria and yeast, eliminating pathogens. Kefir and acidopilus milk lowers cholesterol levels, maintains blood pressure and also proven to check diabetes.

Probiotic fermented non dairy foods are also available *viz.* kimchi- Korean fermented vegetables, fermented fruits, sauerkraut, meat, dry sausages, miso, olives, Indian dishes-idli, dosa, jalebi, pickles, Nigerian dishes-ogi, kunnu, wara and so on. Kimchi is a traditional Korean food manufactured by fermenting vegetables with probiotic lactic acid bacteria (LAB). Many bacteria are involved in the fermentation of kimchi, but LAB become dominant while the putrefactive bacteria are suppressed during salting of cabbage and the fermentation. The addition of other subingredients and formation of fermentation byproducts of LAB promote the fermentation process of LAB to eventually lead to eradication of putrefactive- and pathogenic bacteria, and also increase the functionalities of kimchi. Accordingly, kimchi can be considered a vegetable probiotic food that contributes health benefits in a similar manner as yogurt as a dairy probiotic food. Health functionality of kimchi includes anticancer, antiobesity, anticonstipation, colorectal health promotion, probiotic properties, cholesterol reduction, fibrolytic effect, antioxidative and antiaging properties, brain health promotion, immune promotion, and skin health promotion. Another important food sector where probiotics are extensively used is meat industry. The addition of probiotics to the fermented sausages could promote the health benefits associated with lactic acid bacteria and contribute to the increase in the consumption of such products. The use of probiotics seems more promising in raw fermented meat products like salami as they are made with raw meat and consumed without prior heating, which would kill the probiotic bacteria. Probiotics in meat products give distinct color by nitrate reduction, reducing pH, scavenging oxygen and by degrading hydrogen peroxide. In addition, acid production, proteolysis, lipolysis, delayed rancidity, increased antioxidative properties confers good aroma to meat products. Moreover, low pH, nitrogen reduction and restricting the number of undesirable microorganisms help in preservation. Steps involved in selection and upto commercialisation of probiotics are shown in Fig 2.

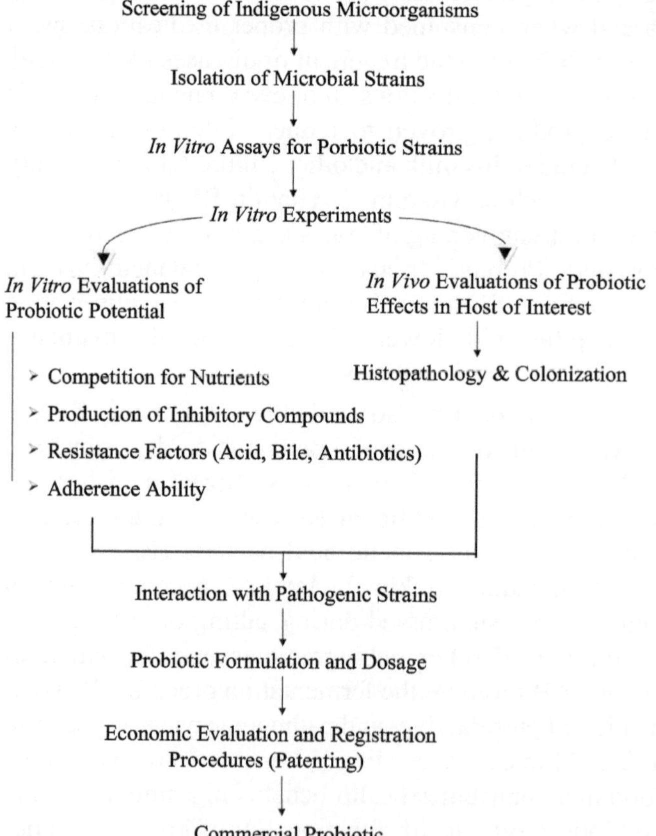

Fig. 2. Steps in the selection of probiotics for animal and human foods.

Nowadays, probiotic bacteria are available in a variety of food products, dietary supplements and drugs for both man and farm animals. As investigations continued in the probiotic field, its concept has been expanded to include bacteria from intestinal origin beside those bacteria isolated from fermented dairy and non dairy products. Since ancient times, man has made and eaten probiotic foods. Although the concept of probiotic bacteria is not new, and is associated with the consumption of fermented foods by human beings, for thousands of years, recently, new types of food products containing probiotic bacteria started to be introduced into the markets, including nondairy products, such as chocolate, cereals, beverages, fruits and vegetables products. In the near future wide range of nontraditional food products containing probiotic bacteria are expected to be introduced into the markets, as the researches in probiotic products development continue in both scientific and commercial centers around the world.

Key questions

1. The bacterial spp which is more efficient in converting organic matter back to CO_2 in pond culture is..............
   a) Salmonella
   b) Bacillus
   c) Staphylococcus
   d) Pseudomonas
2. What is DFM?
3. What is the use of probiotics silage in farm animals?
4. Selection criteria of probiotics for human use

Further reading

Anadón, A., Ares, I., Martínez-Larrañaga, M. R., & Martínez, M. A. (2019). Prebiotics and probiotics in feed and animal health. Nutraceuticals in veter-inary medicine, 261-285.

Chapman, J., Ismail, A. E., & Dinu, C. Z. (2018). Industrial applications of enzymes: Recent advances, techniques, and outlooks. Catalysts, 8(6), 238.

Danfeng Song, Salam Ibrahim and Saeed Hayek. 2012. Recent application of Probiotics in Food and Agricultural Science- Probiotics. DOI10.5772/3444, 656.

Harimurti, S., & Hadisaputro, W. (2015). Probiotics in poultry. Beneficial microorganisms in agriculture, aquaculture and other areas, 1-19.

Shokryazdan, P., Faseleh Jahromi, M., Liang, J. B., & Ho, Y. W. (2017). Probiotics: from isolation to application. Journal of the American College of Nutrition, 36(8), 666-676.

Tshisikhawe, M. L., Diale, M. O., Abrahams, A. M., & Serepa-Dlamini, M. H. (2023). Screening and Production of Industrially Relevant Enzymes by Bacillus paranthracis Strain MHDS3, a Potential Probiotic. Fermentation, 9(11), 938.

Yadav, R., & Shukla, P. (2017). Probiotics for human health: current progress and applications. Recent advances in applied microbiology, 133-147.

9

Diversity of Nisin Producing Lactic Acid Bacteria from Fermented Foods

Introduction

Lactic acid bacteria (LAB) are characterized as Gram-positive cocci or rods, non-aerobic but aerotolerant, able to ferment carbohydrates for production of energy and lactic acid. Lactic acid bacteria includes different major genera: *Lactococcus*, *Enterococcus*, *Lactosphaera, euconostoc, Melissococcus, Oenococcus, Pediococcus, Streptococcus, Tetragenococcus, Vagococcus* and *Weissella*. Other genera are *Aerococcus, Microbacterium*, *Propionibacterium and Bifidobacterium.* Through carbohydrate fermentation along with lactic acid, various organic acids are produced due to which there is a fall in pH level and creates unfavorable environment or inhibits growth of some undesired microorganisms. The low pH makes organic acids lipo -soluble, allowing them to invade into cytoplasm of pathogens by rupturing cell membrane. LAB are generally regarded as safe (*GRAS*), and have a significant role in the preservation of foods and fermented products. Under controlled conditions they can be used as natural competitive microbiota or as specific starter cultures. Antagonistic substance like bacteriocins are produced from lactic acid bacteria that inhibits the activity and growth of other bacteria and these bacteriocins are active against various pathogens in minimum concentration. These are antimicrobial peptides modified by post -translational mechanisms and exported to the extracellular medium. Till date about 239 bacteriocin producing LAB genome and 789 putative bacteriocin producing genes cluster are stored in the public databases. Many gram positive and gram negative bacteria in the environment produce bacteriocins.

Among them the bacteriocin produced from lactic acid bacteria are preferentially used in food industries because of its *GRAS* status, it is non-toxic to eukaryotic cells and they do not interact with sensory quality of the food. Secretion of bacteriocins are based on some environmental factors like bacterial cell density, nutrient availability, the presence of acetic acid, and signalling peptides. Bacteriocins are similar to antibiotics but not actually they are, the major difference between them is that bacteriocin restrict their activity to strains of species related to the producing species and particularly to strains of the same species, whereas antibiotic's activity is in broad spectrum and even if their activity is restricted, this does not show any preferential effect on closely related strains. In addition, bacteriocin are produced during the primary phase of growth, where as antibiotics are secondary metabolites. Generally, bacteriocins are less temperature-labile compared to antibiotics and can withstand extreme pH. Bacteriocins are small cationic molecules of about 30–60 amino acids, forming amphiphilic helices and stable at 100 °C for 10 min and they differ in spectrum of activity, mode of action, molecular weight (MW), genetic origin and biochemical properties. Bacteriocin- producing LAB strains protect themselves from their own toxins by the expression of some specific immune proteins, encoded in the bacteriocin operon. Numerous bacteria can produce bacteriocins and are generally classified based on structures, modes of action, biochemical properties, activity spectra, and target cell receptors.

Bacteriocins produced from Gram positive bacteria classified based on their biochemical and genetic characteristics or the presence of disulfide or monosulfide bonds, molecular weight, thermal stability, proteolytic enzymatic stability, presence or absence of post-translational modification of amino acids, and antimicrobial action. Since from many years in food industries chemical preservatives were used, but there were many health associated risk factor was prevalent. Concern with this issue, food biopreservation techniques used bacteriocins producing LAB becomes a very attractive alternative to chemical preservatives due their bactericidal effect against many food spoilage and pathogenic bacteria without any toxic effect to human health.

The physical parameters influencing the production of bacteriocin are fermentation time, growth kinetics, growth medium composition and pH. Based on the influence of these physical parameters LAB produce varieties of bacteriocins at different percentage. For instance *L. paracasei subsp tolerans FX-6*, *Lactococcus lactis* subsp lactis *A164* produce Nisin-like bacteriocin, *Streptococcusbovis* HC5 produce bovicin HC5 and *Enterococcus faecium spp* SD1,SD2, SD3, AD4 produce enterocin SD1,SD2, SD3, SD4.

Applications of Lactic Acid Bacteria

Table 1. Application of lactic acid bacteria as a starter culture in the production of fermented food and dairy products

LAB Starter culture	**Fermented food products**
Dairy Products a) Cheese (Mesophilic starter)	*Lactococcus lactis spp. lactis, L. lactis spp. cremoris, L. Lactis ssp. Lactis var. diacetylactis, Leuc. mesenteroides spp. cremoris*
Cheese (Thermophilic starter)	*S.thermophilus, Lactobacillus delbrueckii spp. bulgaricus, L. helveticus, L. delbrueckii spp. lactis, L. lactis spp. lactis, L. lactis spp. cremoris*
Yoghurt	*L. delbrueckii spp bulgaricus, S. thermophilus, L. rhamnosus, L. acidophilus*
Yakult	*L. casei spp casei*
Acidophilus milk	*L. acidophillus*
Butter and butter milk	*L. lactis spp lactis, L. lactis spp. lactis var. diacetylactis, L. lactis spp. cremoris, Leuc. mesenteroides spp cremoris.*
Kefir	*L. kefir, L. kefiranofacies, L. brevis, L. plantarum, L. paracasei spp. paracasei, L. delbrueckii spp lactis, L. casei*
Fermented meat products Drysausages	*L. sakei, L. plantarum, L. plantarum, L. pentosus, L. casei, Pediococcus pentosus*
Salami Milano	*L. sakei, L. plantarum*
Salame Piacentino	*L. acidophilus, L. helveticus, L.sakei, L. antri, L. cremoris, L. vaginalis, L. brevis, L. panis, L. versmoldensis, L.zeae, L. curvatus, L. paralimentarius, L. frumenti, L. plantarum, L. graminis, L. reuteri*
Greek dry fermented sausages	*L. sakei, L. plantarum, L.curvatus, L. pentosus, L. lactis spp lactis, W. hellenica, W. paramesenteroides, W. viridescens, W. minor*
Chrorizo	*L.brevis, L. curvatus, L. sakei, L. lactis, P. acidilactici, P. pentosaceus, Leuc. mesenteroids*
Fermented fish products Thaifish	*L. plantarum, L. reuteri*
Pickled fruits and vegetables Cabbage (Sauerkraut)	*Leuc. mesenteroides, L. plantarum, L. brevis, L. fermentum*
Cucumber	*L. plantarum, L. fermentum, L. pentosus, L. acidophilus*
Olives	*Leuc. mesenteroides, L. plantarum, L. brevis, L. fermentum*
Fermented cereal products: Sourdough, Kimchi, Bushera	*L. brevis, L. hilgardii L. sanfransiscensis, L. farciminis, L. fermentum, L. brevis, L. plantarum, L. amylovorus, L. reuteri, L. pontis, L.panis, L. alimentarius, W.cibaria*

Bacteriocin

Bacteriocins are low molecular weight peptides secreted by the predator bacterial cells to kill sensitive cells present in the same ecosystem competing for food and other nutrients. Exceptionally few bacteriocins along with their native antibacterial property also exhibit additional anti-viral and anti-fungal properties. These are generally produced by Gram positive, Gram negative and archaea bacteria. Bacteriocins from gram positive bacteria especially from lactic acid bacteria have broad industrial applications (Juturu *et al.,* 2018). Bacteriocins consist a huge family of ribosomally synthesized peptides that have antibacterial activity towards closely related strains although there are an increasing number of bacteriocins reported to have broad range antimicrobial activity. In the past decade, interest in bacteriocin research, especially from lactic acid bacteria (LAB), has gained great momentum due to its potential as both a natural food preservative and as therapeutic antibiotics. Bacteriocins have a number of positive attributes that have made them especially attractive for various applications. LAB bacteriocins are inherently tolerant to high thermal stress and are known for their activity over a wide pH range. Bacteriocins have many positive attributes that have made their existence for various applications. LAB bacteriocins are tolerant to high thermal stress and active at wide pH range. These antimicrobial peptides are also colourless, odourless, and tasteless, which further enhance their potential usefulness.

For food applications, LAB bacteriocins have vast advantage over conventional antibiotics and LAB bacteriocins are generally considered food-grade due to the typical association of LAB to food fermentation that dates back to ancient times. In fact, the U.S. Food and Drug Administration (FDA)classified LAB and its by-products as Generally Regarded as Safe (GRAS) as a human food ingredient. Whereas for clinical applications, bacteriocins have been presented as a viable alternative to antibiotics due to the high specificity of some bacteriocins against clinical pathogens, including multi-antibiotic resistant (MDR) strains (Cotter *et al.,* 2013).

Among bacteria, production of bacteriocin is widespread. Bacteriocins are synthesized by majority of bacterial species. Because of simple bio synthetic machineries and are often associated with transferable elements such as conjugative transposons or plasmids. plamids, as highlighted earlier, bacterocins are, bacteriocins are ribosomally synthesized peptides. Genes related to bacteriocin biosynthesis are generally clustered, and are encoded on plasmids, chromosome and/ or transposons with minimum genetic machinery consisting of structural cognate immunity genes. Bacteriocins are usually synthesized as biologically inactive prepeptides that include an N-terminal leader peptide attached to the C-terminal propeptide.

Types of Bacteriocin

Bacteriocins from gram negative bacteria

There are mainly four classes of bacteriocins produced from gram negative bacteria: colicins, colicin-like, phage-tail-like bacteriocins, and microcins. Colicins are protease-sensitive, heat-sensitive high-molecular weight of about 30-80 kDa. Bactericidal proteins synthesized by most *E. coli* strains that have a single colicinogenic plasmid. Compounds of this class are the frequently studied and are used as model systems for studying the structures, functions, and evolution of bacteriocins. Colicin synthesis is carried out under stress and is fatal for the producing cells, due to co-expression with lysis protein. Depending on the mechanism of interaction with the target cell, they are divided into three main groups-pore-forming, nuclease, and peptidoglycan degrading.

Phage tail-like bacteriocins also known as tailocins are larger protein structures with size of 20–100 kDa consisting of eight to fourteen different polypeptide subunits that show similarities to the bacteriophage tail modules. Tailocins are encoded in the genomes of bacteria by a cluster of genes larger than 40 kbp. The locus includes genes encoding structural proteins, assembly enzymes, chaperones, regulatory genes, and lysis cassettes, their function is to release bacteriocins into the environment. This type of bacteriocins are divided into two groups-R and F.

R-type tailocins are evolutionarily related to the tails of phages in the family of *Myoviridae* and forms a long shell-encircled tube, at one end of which is a complex basal plate with receptor-binding proteins (RBP). F-type bacteriocins belonging to the tails of phages in the *Siphoviridae* family do not have a shell. The mechanism of action of tailocins is not fully understood, and presumably involves compression of the shell and penetration of the nucleus through the cell wall, which leads to the formation of a channel or pore that violates the membrane potential of the target cell. The induction of synthesis of these compounds is associated with the SOS response of the producer cell to DNA damage.

Microcins are the third type of bacteriocins produced by Gram negative bacteria of size less than 10kDa and these are highly stable peptides that exhibit competitive interactions between members of the *Enterobacteriaceae family*. Microcins are resistant to proteases, extreme pH and temperature values. Clusters of genes encoding microcins are located in plasmids, and often less found in genomic DNA. Variable number of genes are included, but demonstrate a conservative organization: open reading frames encoding a microcin precursor, secretion proteins, immune factors, and, in some cases, post-translational modification enzymes. Microcins are divided as Class I and Class II based on the presence, nature, and localization of modifications, as well as the gene cluster organization and the sequence of leader peptides.

Class I have peptides with a molecular weight below 5 kDa (microcins B17, C7-C51, and J25) with posttranslational modifications. Bacteriocins of this group inhibit vital bacterial enzymes, such as DNA gyrases I and II and RNA polymerases, and can also have an inhibitory effect on the respiratory chain.

Class II includes larger (5–10 kDa) plasmid peptides without post-translational modifications (MccL, MccV, and Mcc24) and linear microcins encoded in the chromosome and carrying a C-terminal siderophore (microcins E492, M, H47, I47, and G47). Their toxic effect is in the formation of pores and destruction of the target cell membrane.

Bacteriocins from Gram positive bacteria

Gram positive bacteriocins are mainly classified into three classes based on the biochemical and genetic characteristics. Classes I and II have 2 and 3 sub classes, respectively. Modern classification based on the structural similarity, phylogenetic evolution and consensus motif sequence, bacteriocins are classified into 12 groups. Groups 1, 3 and 4 are further sub-divided into sub groups a and b. There are wide number of bacteriocins produced by different LAB and they can be classified according to their biochemical and genetic characteristics (Klaenhammer., *et al.,* 1994; González-Martínez., *et al.,* 2003).

ClassI: The class I bacteriocin include lantibiotics: these are small in size of about less than 5kDa, heat-stable peptides, acts on membrane structures and modified by post-translational modifications. These arise *via* a two-step process, originated from post translational modifications: firstly, gene-encoded serine and threonine are subjected to enzymatic dehydration to give rise to dehydroalanine and dehydrobutyrine, respectively. A very well-known example of this group is nisin. The lantibiotic bacteriocins were initially divided into two subclasses, based on structural similarities.

Subclass Ia consists relatively elongated, flexible and positively charged peptides; they generally act by formation of pores in the cytoplasmatic membranes of sensitive target species. The prototypic lantibiotic nisin is a member of this group.

Subclass Ib peptides are characteristically globular, more rigid in structure and are either negatively charged or have no net charge. They exert their action by interfering with essential enzymatic reactions of sensitive bacteria. For instance class I bacteriocin are Mersacidin and Nisin.

The major pathways for Class I (lantibiotic) bacteriocin biosynthesis can be explained by using the pathway followed by the well-known nisin, although there are slight differences for the non-lantibiotic bacteriocins as they do not need incorporation of unusual amino acids.

1. The *nisA* gene is translated to pre-nisinA peptide;
2. Pre-nisin A is converted into precursor nisin A by the products of the *nisB* and *nisC* genes. Several disulphide bridges are made and some amino acids are converted to unusual ones.
3. The precursor nisin A is exported out of the cell with the aid of the *nisT* and *nisP* while the leader peptide is simultaneously cleaved to obtain the nisin A.

Bacteriocin expression is regulated either by external induction factors, usually secreted by the producer strain itself, or it depends on environmental conditions such as temperature and pH. In recent years, production of nisin has been significantly improved through genetic modifications to nisin producer strains and through innovative applications in the fermentation process. Recently, 15,400 IU ml^{-1} nisin production has been achieved in *L.lactis* cells following genetic modifications by eliminating the factors that negatively affect nisin biosynthesis or by increasing the cell density of the producing strains in the fermentation medium.

Class II

These are non-Lantibiotics bacteriocins of variable molecular weight of about less than 10kDa. These are heat-stable, have regular amino acids. This group is divided into three subgroups: Class IIa peptides are active against *Listeria*, examples are pediocin PA-I and sakacin P. Pediocin PA-I (Fig. 1) is a broad spectrum bacteriocin produced from LAB that shows a particularly strong activity against *Listeria monocytogenes*, a food borne pathogen of special concern among the food industries and Sakacins P restricts closely related *Lactobacilli, Listeria monocytogenes* and *Ivanovvi, Enterocccus, Corynibacterium spp* and *Brocothrix thermosphacta.*

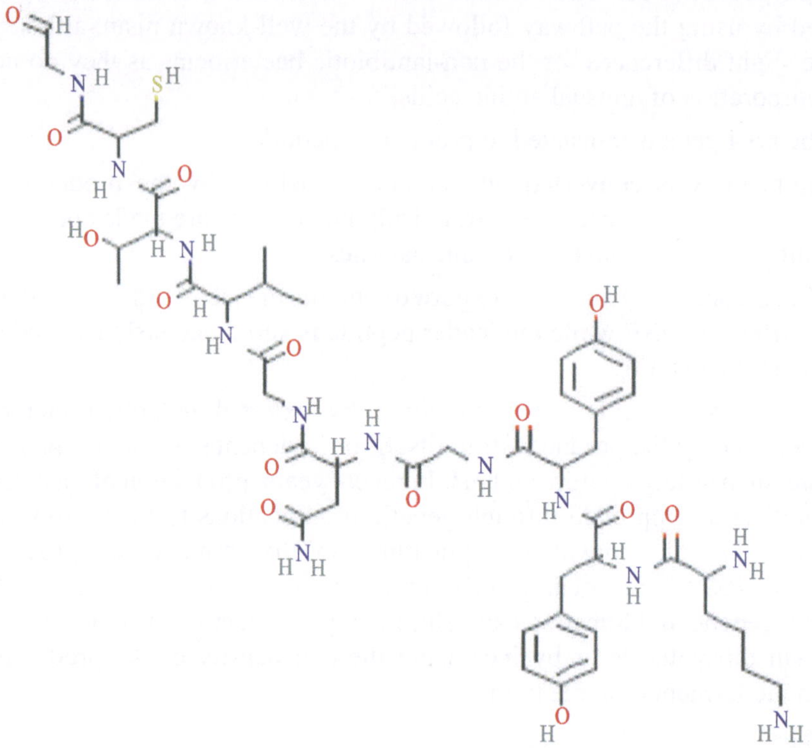

Fig. 1. Structure of Pediocin A obtained from Pubchem (NIH).

Class IIb are formed by a complex of two distinct peptides. These peptides have minute or no activity and it has no sequence similarities between complementary peptides. Plantaricin EF, Plantaricin JK and Lactocin-F are the examples. Lactocin-F is a non lantibiotic, heat-stable, peptide bacteriocin produced by lactobacillus. Molecular analysis of the Lactocin-F DNA region characterized a small operon that codes for three operon reading frames, named as lafA, laf X, and ORF. Plantariocins are proteinaceous or peptidic toxins produced by bacteria to inhibit the growth of similar or closely related bacterial strains (Fig. 2).

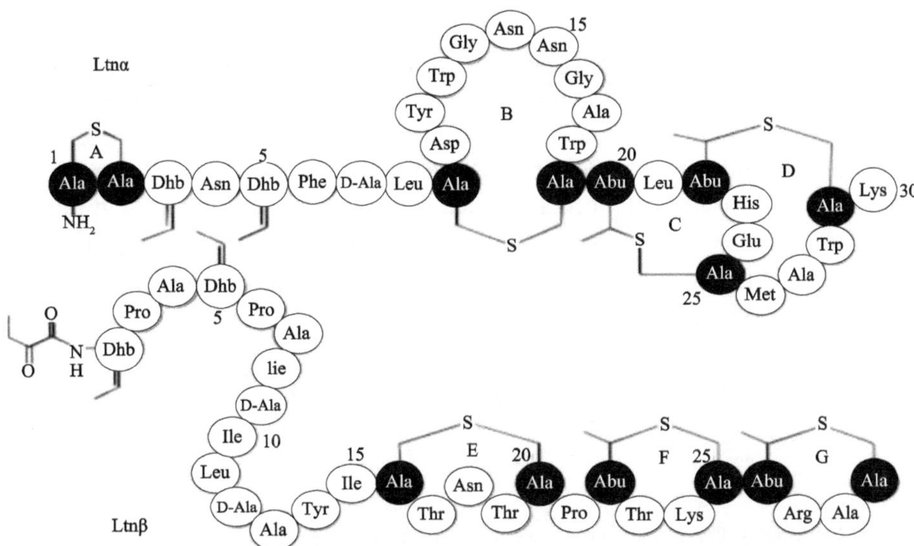

Fig. 2. Molecular Structure of Lacticin 3147 (Ltn α, Ltnβ) (Jhon and Deseenthum., 2015).

Class IIc bacteriocins are small peptides, heat-stable, which are transported by leader-peptides. Divergicin A and acidocin B falls in this subclass produced by *Lactobacillus acidophilus* M46, was originally reported to be a linear peptide composed of 59 amino acid residues. However, its similar to the bacteriocin Gassericin A, produced from *Lactobacillus gasseri LA39.* Sakacins P inhibits closely *related Lactobacilli, Listeria monocytogenes and Ivinovvi, Enterociccis, Corynibacterium spp and Brocothrix thermosphacta.*

Class IIc

Bacteriocins under this subclass are large peptides, with molecular weight that exceeds 30 kDa. Helveticins J, acidofilicin A and lactacins A and B are examples of Class IIc bacteriocins. Many low molecular weight bacteriocins are highly cationic falls at pH 7.0 and this seems to be a unifying feature of both the lantibiotics and non lantibiotics. Lantibiotics are the most studied and explored industrially because of their vast applications.

CLASS III

Bacteriocin belongs to the class III have peptides with molecular weight of 6.3 kDa. These are heat-stable, detected only in cultures maintained between pH 5.0 to 6.0. Acidophilucin A from *Lactobacullus acidophilus* LAPT1060, act as antimicrobial agent and active against six strains of *Lactobacillus delbrueckii* subsp *bulgaricus* and also act against six strains of *Lactobacillus helveticus*.

CLASS IV

Oscariz and Pisabarro (2001) described fourth classes of bacteriocins with sub class. The fourth class of bacteriocins are large protein and contains mixture of bacteriocins with other macromolecules. Example of Class IV bacteriocins are Leucin S and Mesenterocin 52, whereas Leucin S is used in the making of syrup which is used to treat leucorrhoea. Mesenterocin 52 bacterocin inhibits *Leuconostoc* strains and several strains of *Enterococus* & *Listeria sp.* (Fig. 3).

Among the different bacteriocins, Class I and II are more applicable in food industry due to their specific activity and robustness.

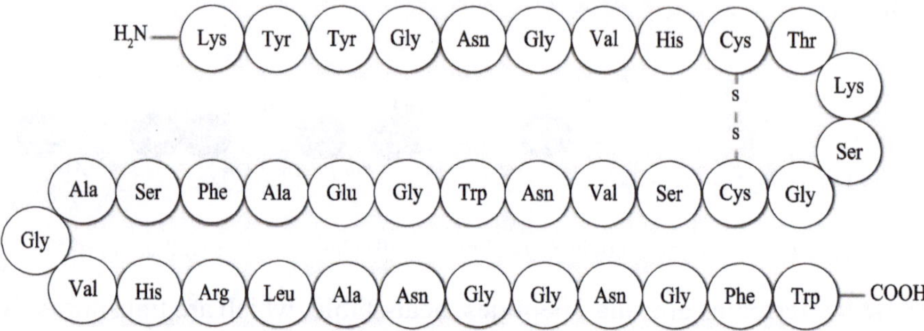

Fig. 3. Molecular structure of Leucocin produced from Leuconostocmesenteroides (Lohans and vedorans., 2012)

Nisin

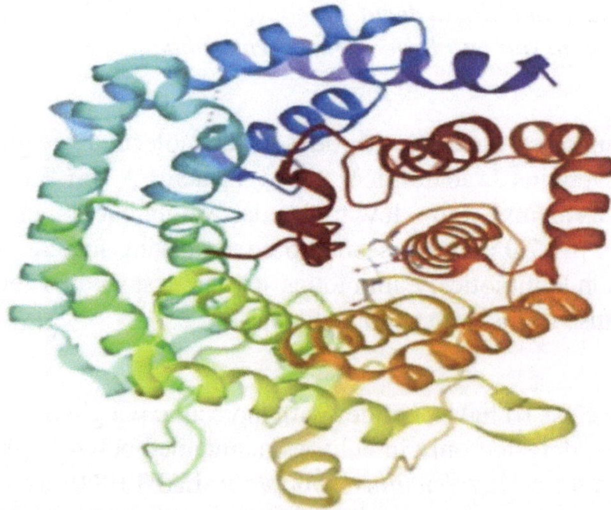

Fig. 4. 3D structure of Nisin

"Do you know?....
In 2023, for the first
time purification of
commercial
nisin solution was done
using electrodialysis"

NISIN: A bacteriocin Nisin belongs to type I lantibiotic group, that produced from *Lactococcus lactis*. This bacteriocin has a wide antimicrobial activity against the spores of Gram positive bacteria, and also act against the food pathogens mainly includes *Listeria monocytogenes, Staphylococcus aureus, Clostridium* and *Bacillus species.* Also it act against gram negative pathogens that includes *Escherichia coli* and *Salmonella.* In 1969, Food and Agriculture Organisation from World Health Organization (FAO/WHO) expert committee considered nisin, as safe for use in food products. Later in 1983, it was considered and added in the European list of food additives and in 1988,this was approved by the US Food and Drug Agency (FDA) as generally regarded as safe (GRAS) for use in pasteurized products and processed cheeses to inhibit the growth of *Clostridium botulinum.* Nisin consists of 34 amino acid residues, with molecular weight of 3.5 kDa, and isoleucine (NH_2) and lysine (COOH) as terminal amino acids (Fig. 5).

Structure of Nisin

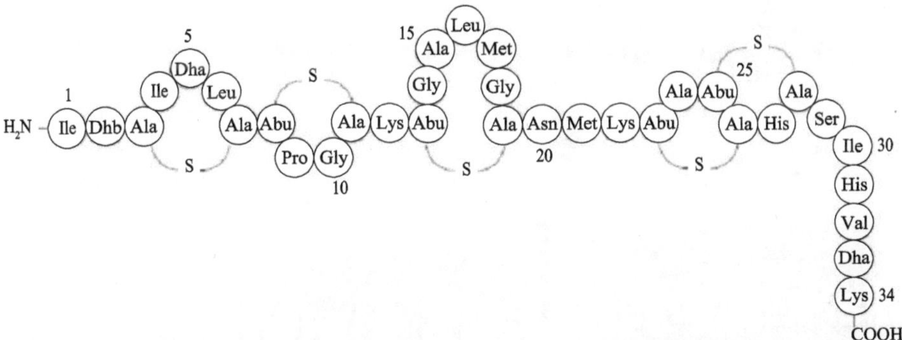

Fig. 5. Schematic representation of the primary structure of nisin produced by *L. lactis subsp. lactis* ATCC 11454 represents lanthionine, Abu-S-Ala-β-methyl, lanthionine, dehydroalanine Dha, and Dhbdehyrobutyrine (Jozala *et al.*, 2014).

Nisin was first used during production of cheese. It has an immune modulator role and a selective cytotoxic effect against cancer cells. However, it is notable that the high production cost of nisin is an important issue that faces difficulties in the use. The major reason is the low nisin production yield of producer strains. However, it is notable that the high production cost of nisin is an important issue which restricts its intensively use. The major reason is the low nisin production yield of producer strains. Hence the production of nisin has been improved by genetic modifications in particular strains of micro-organisms that produce nisin. About 15400 IU/mL nisin production has been successfully done using *L. lactis* cells increasing the density of the producing strains in the fermentation medium or by genetic modifications with eliminating the factors that affect nisin biosynthesis. It has an immune modulator role and a selective cytotoxic effect against cancer cells.

There are two variants of this bacteriocin: nisin A and Z, which differ from each other only by the amino-acid. Nisin A type has Histidine as amino acid and Nisin Z has Asparagin as amino acid. Predominantly nisin is used in canned foods and dairy products and is very effective when utilized in the production of processed cheese and cheese spreads acting against heat-resistant spore forming micro-organisms such as *Bacillus* and *Clostridium*. It is also effective against Gram-positive mastitis-causing pathogens. It has a great potential for use with medical purposes as it is active under physiological pH. It also act against human pathogens as methicilin-resistant *Staphylococcus aureus* (MRSA), vancomycin resistant *Enterococcus faecalis*, penicillin-resistant *Pneumococcus, Propionibacterium acne* and *Streptococcus mutans*.

Mechanism of action of nisin

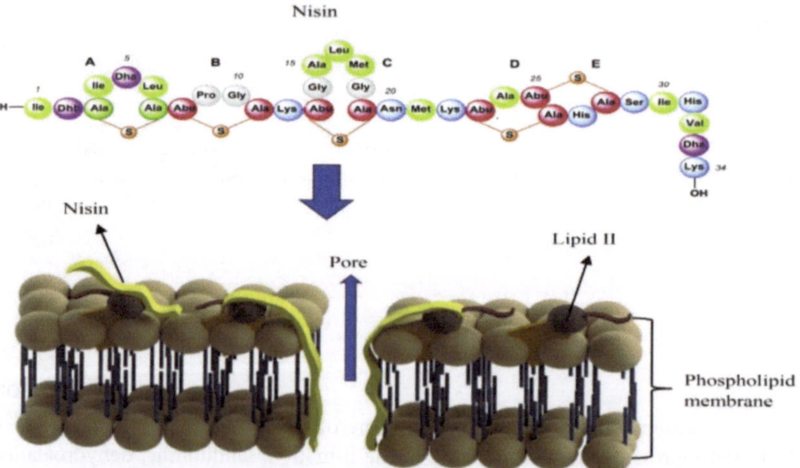

Fig. 6. Antimicrobial mechanism of nisin action to the killing of bacteria

The mechanism of lantibiotic action is most extensively studied for nisin (Fig. 6).

There are two structural domains, one at the N-terminus contains three (b-methyl) lanthionine rings namely A, B and C. The other domain at the C-terminus contains two rings (D and E), this allows nisin to have a dual action. The presence of rings A, B and C facilitates the binding of nisin to pyrophosphate moiety of lipid II peptide, and rings D and E are involved in pore formation. When it is linked to lipid II, it blocks the synthesis of cell walls and in some cases, may use lipid II to secure itself in the membrane and to initiate pore formation. Figure-6 shows the suggested mechanism of nisin action. At the initial stage, nisin binds to the pyro phosphate of lipid II. Then the C-terminus of the molecule is transfered through the membrane and appears on its inner side. At this stage, clusters A, B, C, D and E in the nisin molecule is formed. The interaction with lipid II stabilizes the transmembrane orientation of the peptide, which results in the formation of a stable pore of 2 nm in diameter. Eight molecules of nisin and four molecules of lipid II are involved in the formation of one pore. For specific membrane binding class I bacteriocins use lipid II. The action mechanism of lantibiotics with similar structures of A and B rings depends on lipid II. In particular, mersacidin and actagardine (type B lantibiotics) also form a complex with lipid II, but this binding only blocks the incorporation of lipid II into peptidoglycan, which leads to cell lysis slowly. At the level of transglyco-sylation, mersacidin blocks the peptidoglycan synthesis but in this case pores are not formed in the membrane.

Nisin variants

Many variants of nisin producing microorganisms have been found in environment. Nisin A was first discovered in *L. lactis*, this organism commonly found in dairy products and it is most widely studied nisin variant. As antimicrobials, Nisin A and Z share similar properties, but nisin Z has a superior rate of diffusion and solubility under neutral pH conditions. Nisin F is another type of nisin can be isolated from *L. lactis* F10 mainly from faeces of a freshwater catfish in South Africa and differs from nisin A by two amino acid residues. Nisin F has two amino acid substitutions at position 27 and 30. Nisin Q can be isolated from the bacteria *L. lactis* abundant in river situated in Japan. Nisin Q contains four substitutions at positions 15, 21, 27 and 30. Nisin A, Z, F and Q have effective antimicrobial activity against Staphylococcus aureus. More distantly related variants Nisin U and U2 are found to be isolated from *Streptococcus uberis*, an organism that commonly inhabits the lips, skin and udder tissues of cows and is found in raw milk. Nisin U and U2 contain nine and ten amino acid substitutions, respectively, compared to nisin A. From the recent studies it is found that nisin H can be isolated from a *Streptococcus hyointestinalis* strain derived from porcine intestine.

Nanotechnology and food integration

When nisin is used in food products it is influenced and depends on various factors. Efficacy of nisin can be reduced due to the binding or interaction of nisin with food matrix components and hence it also decreases the stability of food. Several food components, such as glutathione, proteases, sodium metabisulfite, and titanium dioxide can severely affect action of nisin. Its effectiveness is reduced as it is free form as it can be degraded by enzymes. Also a positive charge that resides on the bacterial cell wall reduces effectiveness of nisin by preventing the interaction of nisin and bacterial surface molecules due to electrostatic repulsion. Hence use of nanoparticulate systems improves controlled release and delivery of nisin. Few of the functional ingredients of food with varying molecular and physical structures are antioxidants, antimicrobials, flavoring, colorants and food preservatives. They are usually applied as coatings or incorporated into some form of delivery system instead of utilizing in their pure form. Hence, several nanotech-based delivery systems have been formulated, including nanoliposomes, nano-emulsions, nanoparticles and nanofibers.

Applications of nisin

In food industry

Many researchers around the world have shown the promising potential and activity of nisin that act against growth of pathogens and also controlling potential in both dairy and non-dairy food products. The use of nisin has been widely applied in the production of cheese at low pH. The earliest application of nisin in dairy products was studied to the prevent the spoilage caused by *Clostridial* species that responsible for the late-blowing defect in cheese. Rodríguez *et al.* (2000) described the effect of two nisin- producing strains, *L. lactis subsp. lactis TAB 50* and its mutant strain TAB 50-M4, in the production of cheese from raw milk inoculated with *S. aureus*, and found that they acted against *S. aureus*. The factors influencing the use of nisin in dairy products should be well defined before the use for better implementation of the techniques.

Infectious Diseases

Antibiotic-resistant especially biofilm-associated infections are difficult to prevent or treat. With the over usage of antibiotics and antiviral drugs, drug resistance that made a way to search for alternative methods to treat. Some scientists like Piper and co-workers reported that the effective antibacterial activity of nisin against antibiotic resistant *Staphylococci.* Some of the *in vitro* effects of nisin against MRSA strains, were also studied by the scientist Dosler and their collegues and also confirmed that nisin is a appropriate bacteriocin to combine with conventional antibiotics, such as vancomycin, or ciprofloxacin. Dosler and colleagues investigated the *in vitro* effects of nisin against MRSA strains, and concluded that nisin was a good candidate for further research. Other Nisin-based formulation are also effective in the treatment of clinical mastitis in lactating dairy cows caused by several different mastitis pathogens. In treatment of sub clinical by multiple mastitis pathogens in lactating dairy cows, nisin Z was effective.

Nisin as a cancer Therapeutic agent

Many scientists explored the cytotoxic and antitumor characters of nisin A and discovered that it would reduce squamous cell carcinoma tumorigenesis. It has been investigated that the translational potential of a high purity form of nisin Z. In humans FDA has approved nisin of about 83.25 mg/kg as the no-observed-effect-level (NOEL), and 66.7mg/kg in mice as a cancer therapeutic dose. In addition, a combination of conventional cancer drug, doxorubicin with nisin can regulate in terms of decreasing tumour severity in skin carcinogenesis.

Nisin, as modulator of gut microbiota

Several recent studies provide evidence and confirms that bacteriocins and bacteriocin-producing bacteria can be used for modification of gut microbiota, making them an attractive strategy to address gut-related diseases and disorders. Some of the issues associated with conventional antibiotics can be addressed by bacteriocins. To influence the gut microbiota or to control specific pathogens that are associated with chronic intestinal diseases the efficiency of nisin was investigated that it can be either applied directly or secreted by *in-situ* bacteria. For instance, it was evident when nisin was supplemented in poultry diet, it can modulate the gastrointestinal ecology and improved growth performance in a benficial way. Similar effects were aslo studied in rabbit models. Therefore, nisin treatments elicited significant changes in the gastrointestinal microbiota by increasing benficial microbes such as *Lactobacillus, Bacteroides, and Bifidobacterium* by reducing pathogenic strains of *E.coli* and *Enterococcus spp.* Nisin also targets the gut pathogen *C. difficile* hence it also been considered for its therapeutic potential. Indiscreminate usage of antibiotics allowed that allow *C.difficile* to proliferate and cause recurrent *C.difficile*-associated diarrhoea and intestinal inflammatory disease (collectively designated *C.difficile* infection or CDI). Therefore usage of nisin is regarded as safe and perhaps useful against *C. difficile* infection.

Antimicrobial Packaging Using Nisin

To overcome nisin limitations and for developing antimicrobial packaging, incorporation of nisin into the packaging material was one possible stratergy. By this the interaction of antimicrobials with food components (mainly protein and fat), dilution into bulk foods, neutralization, and rapid diffusion in the food mass can be reduced. Nisin binds to different compounds as it is highly surface-active molecule that creates suitable environment for adsorption on solid surfaces and killing bacterial cells that sub sequently adhere. Hence, these nisin activated surfaces were found to be potential for applications as antimicrobial packaging. To enhance the stability bioactivity in a food environment and to ensure its controlled release to the dairy product one of the promising method is encapsulation of nisin to the packaging. This was confirmed by inhibiting *L. monocytogenes* on Cheddar cheese without affecting or altering its sensory properties.

Key questions

1. Amino acid present in Nisin A type bacteriocin-

   a) Thyrosine b) Methionine

   c) Histidine d) Aspargine

2. Mesenteroicin belongs to -

   a) Class II bacteriocin b) Class IV bacteriocin

   c) Class IIc bacteriocin d) Class III bacteriocin

3. Nisin F has two amino acid substitutions at:

   a) Postion 27 and 30 b) Position 30 and 25

   c) Position 22 and 32 d) Position 25 and 30

4. Lanthionine rings in nisin are

   a) A, B and D b) A, B and C

   c) D and E d) B, D and E

5. Dozler concluded that nisin was combined with conventional antibiotics such as:

   a) Vancomycin b) Ciprofloxacin

   c) Both a and b d) Only a

6. Write a short notes on:
   i) Mechanism of action of nisin and its variants. Explain
   ii) Describe the types of bacteriocins.
   iii) Differentiate antibiotics and bacteriocins.
   iv) Elaborate the applications of nisin.

Further reading

Alvarez-Sieiro, P., Montalbán-López, M., Mu, D., &Kuipers, O. P. (2016). Bacteriocins of lactic acid bacteria: extending the family. Applied microbiology and biotechnology, 100, 2939-2951.

Bernela, M., Kaur, P., Chopra, M., & Thakur, R. (2014). Synthesis, characterization of nisin loaded alginate–chitosan–pluronic composite nanoparticles and evaluation against microbes. LWT-Food Science and Technology, 59(2), 1093-1099.

Bintsis, T., &Athanasoulas, A. (2015). Dairy starter cultures. Dairy Microbiology, a Practical Approach; Papademas, P., Ed.; CRC Press: Boca Raton, FL, USA, 114-154.

Cao, L. T., Wu, J. Q., Xie, F., Hu, S. H., & Mo, Y. (2007). Efficacy of nisin in treatment of clinical mastitis in lactating dairy cows. Journal of dairy science, 90(8), 3980-3985.

Carr, F. J., Chill, D., & Maida, N. (2002). The lactic acid bacteria: a literature survey. Critical reviews in microbiology, 28(4), 281-370.

Cintas, L. M., Casaus, M. P., Herranz, C., Nes, I. F., & Hernández, P. E. (2001). Bacteriocins of lactic acid bacteria. Food Science and Technology International, 7(4), 281-305.

Cui, H. Y., Wu, J., Li, C. Z., & Lin, L. (2016). Anti-listeria effects of chitosan-coated nisin-silica liposome on Cheddar cheese. Journal of dairy science, 99(11), 8598-8606.

Deegan, L. H., Cotter, P. D., Hill, C., & Ross, P. (2006). Bacteriocins: biological tools for bio-preservation and shelf-life extension. International dairy journal, 16(9), 1058-1071.

Djadouni, F., &Kihal, M. (2012). Antimicrobial activity of lactic acid bacteria and the spectrum of their biopeptides against spoiling germs in foods. Brazilian Archives of Biology and Technology, 55, 435-444.

Dosler, S., &Gerceker, A. A. (2012). In vitro activities of nisin alone or in combination with vancomycin and ciprofloxacin against methicillin-resistant and methicillin-susceptible Staphylococcus aureus strains. Chemotherapy, 57(6), 511-516.

Field, D., Fernandez de Ullivarri, M., Ross, R. P., & Hill, C. (2023). After a century of nisin research-where are we now?. FEMS microbiology reviews, 47(3), fuad023.

Galvin, M., Hill, C., & Ross, R. P. (1999). Lacticin 3147 displays activity in buffer against Gram□positive bacterial pathogens which appear insensitive in standard plate assays. Letters in Applied Microbiology, 28(5), 355-358.

Haller, D., Colbus, H., Gänzle, M. G., Scherenbacher, P., Bode, C., &Hammes, W. P. (2001). Metabolic and functional properties of lactic acid bacteria in the gastro-intestinal ecosystem: a comparative in vitro studybetween bacteria of intestinal and fermented food origin. Systematic and Applied Microbiology, 24(2), 218-226.

Hasper, Hester E., Naomi E. Kramer, James L. Smith, J. D. Hillman, Cherian Zachariah, Oscar P. Kuipers, Ben De Kruijff, and EefjanBreukink. "An alternative bactericidal mechanism of action for lantibiotic peptides that target lipid II." Science 313, no. 5793 (2006): 1636-1637.

Hsu, S. T. D., Breukink, E., Tischenko, E., Lutters, M. A., De Kruijff, B., Kaptein, R., ...& Van Nuland, N. A. (2004). The nisin–lipid II complex reveals a pyrophosphate cage that provides a blueprint for novel antibiotics. Nature structural & molecular biology, 11(10), 963-967.

Joerger, M. C., &Klaenhammer, T. R. (1986). Characterization and purification of helveticin J and evidence for a chromosomally determined bacteriocin produced by Lactobacillus helveticus 481. Journal of bacteriology, 167(2), 439-446.

Jolly, L., Vincent, S. J., Duboc, P., &Neeser, J. R. (2002). Exploiting exopolysaccharides from lactic acid bacteria. In Lactic Acid Bacteria: Genetics, Metabolism and Applications: Proceedings of the seventh Symposium on lactic acid bacteria: genetics, metabolism and applications, 1–5 September 2002, Egmondaan Zee, the Netherlands (pp. 367-374). Springer Netherlands.

Jozala, A. F., de LencastreNovaes, L. C., & Junior, A. P. (2015). Nisin. In Concepts, compounds and the alternatives of antibacterials. IntechOpen.

Juturu, V., & Wu, J. C. (2018). Microbial production of bacteriocins: Latest research

Meira, S. M. M., Zehetmeyer, G., Werner, J. O., &Brandelli, A. (2017). A novel active packaging material based on starch-halloysite nanocomposites incorporating antimicrobial peptides. Food Hydrocolloids, 63, 561-570.

Mokoena, M. P. (2017). Lactic acid bacteria and their bacteriocins: classification, biosynthesis and applications against uropathogens: a mini-review. Molecules, 22(8), 1255.

Moreno, Y., Collado, M. C., Ferrús, M. A., Cobo, J. M., Hernández, E., & Hernández, M. (2006). Viability assessment of lactic acid bacteria in commercial dairy products stored at 4 C using LIVE/DEAD® BacLightTM staining and conventional plate counts. International journal of food science & technology, 41(3), 275-280.

Rodríguez, E., González, B., Gaya, P., Nuñez, M., & Medina, M. (2000). Diversity of bacteriocins produced by lactic acid bacteria isolated from raw milk. International Dairy Journal, 10(1-2), 7-15.

Sahl, H. G., &Bierbaum, G. (1998). Lantibiotics: biosynthesis and biological activities of uniquely modified peptides from gram-positive bacteria. Annual Reviews in Microbiology, 52(1), 41-79.

Shin, J. M., Gwak, J. W., Kamarajan, P., Fenno, J. C., Rickard, A. H., &Kapila, Y. L. (2016). Biomedical applications of nisin. Journal of applied microbiology, 120(6), 1449-1465.

Sobrino-López, A., & Martín-Belloso, O. (2008). Use of nisin and other bacteriocins for preservation of dairy products. International Dairy Journal, 18(4), 329-343.

Suda, S., D Cotter, P., Hill, C., & Paul Ross, R. (2012). Lacticin 3147-biosynthesis, molecular analysis, immunity, bioengineering and applications. Current Protein and Peptide Science, 13(3), 193-204.

Suda, S., D Cotter, P., Hill, C., & Paul Ross, R. (2012). Lacticin 3147-biosynthesis, molecular analysis, immunity, bioengineering and applications. Current Protein and Peptide Science, 13(3), 193-204.

Zimina, M., Babich, O., Prosekov, A., Sukhikh, S., Ivanova, S., Shevchenko, M., &Noskova, S. (2020). Overview of global trends in classification, methods of preparation and application of bacteriocins. Antibiotics, 9(9), 553.

Zouhir, A., Hammami, R., Fliss, I., &Hamida, J. B. (2010). A new structure-based classification of gram-positive bacteriocins. The protein journal, 29, 432-439.

10

New Insights on the Role of Lactic Acid Bacteria (LAB) in Immune Modulation

Introduction

The gut microbiota is the complex and dynamic community of microorganisms that includes various bacteria, fungi and other microorganisms, proliferate and lead a symbiotic/mutual relationship with the host system. These microorganisms protect the host against pathogens by unique mechanisms, helps to maintain shape and strengthen the intestinal epithelium and boost immune response. Diverse bacterial metabolites can immigrate the host bloodstream and other parts of the body *via* various channels in humans and animals, such as gut-microbiota-brain, gut microbiota-skin, gut-vagina, gut-liver, gut-bones etc. However, many aspects of modern lifestyle would alter the gut microbiota and lead to numerous diseases including inflammatory bowel disease, diabetes mellitus, obesity, and metabolic syndrome. Probiotics and other beneficial microbes exert many beneficial effects on their hosts by modulating the gut microbiota. While the term probiotic is reserved only for well-characterised strains with clinically proven health benefits, other beneficial microorganisms include, organisms responsible for the fermentation of foods *viz.* yogurt, kefir, kombucha, kimchi and many others that have been shown to have distinct health benefits. Interestingly, major group of these beneficial organisms belong to lactic acid bacteria.

Lactic acid bacteria (LAB) are ubiquitous, Gram-positive, usually non-motile, catalase-negative, aero-tolerant bacilli or cocci with high tolerance to low pH (lactic acid). Lactic acid bacteria produce variety of metabolites which have majority of therapeutic and functional properties that have a profound impact on human health. In this chapter, we shall discuss about,

- Influence of lactic acid bacteria on gut microbiota composition,
- Bioactive compounds yielded by lactic acid bacteria and their influence on the gut microbiota composition,
- Molecular mechanisms underlying the effects of lactic acid bacteria on immunity, pathogen exclusion, intestinal barrier function related to the gut microbiota and beyond,

- Potential therapeutic interventions with lactic acid bacteria related to the gut microbiota and beyond.

Organic acids like lactic acid (LA) and short-chain fatty acids (SCFA) produced from LAB through carbohydrate fermentation in the intestine. LA is highly hygroscopic aliphatic acid and water-soluble. LAB are divided into two distinct clades on the basis of utilization of homofermentative or heterofermentative metabolism. LA plays an important role in wine making, fermentation of food and also in biopharmaceutical industries. Because of its antimicrobial and immunomodulatory properties, it is used in clinical treatment. LAB have the ability to ferment dietary fibres and sugars to produce different SCFAs. Acetate, butyrate, and propionate are some SCFA produced from LAB. Different types of bacteriocins are secreted by different strains of LAB which have unique mode of action, molecular weight, genetic origin and biochemical properties. They disrupt the membrane vesicles by acting on cytoplasmic membrane, this property of LAB helps to target a specific sensitive pathogen.

Bacteriocins from LAB used as biological preservatives in various food products in food industries, they are evaluated as Generally Recognized as Safe (GRAS) compounds by the FDA. Nisin is a bacteriocin produced from *Lactococcus lactis* strains and widely used in the commercial development of food preservative. It has decomposition capabilities mediated by digestive enzymes and wide antimicrobial spectrum. Other bacteriocins include Leucocin produced from *Leuconostoclactis SM2*, has maximum antimicrobial activity against target bacterial strains. *Carnobacterium maltaromaticum KCA018* produce bacteriocin named KCA, effective in inhibiting Gram-positive bacteria and used as bio-preservative with narrow spectrum antimicrobial activity. Plantaricin Bio-LP1 is a novel potential bacteriocin produced from *Lactiplantibacillus plantarum NWAFU-BIO-BS29.*

LAB based amino acids are known to play a prominent role in human health promotion, for instance, a nonessential amino acid called citrulline used in the treatment of intestinal inflammation. While, an essential amino acid, ornithine has several important physiological functions such as anti-obesity activity, stimulation of hormone growth and promotion of muscle growth. Among different amino acid metabolites of LAB, γ-Aminobutyric Acid (GABA) is a neuro inhibitory molecule. It is a non-protein amino acid widely found in micro-organisms, animals and plants. GABA is broadly used in treatment of psychiatric disorders such as epilepsy, convulsion and Parkinson's disease because of its various benefits effects, as it is inhibitory neurotransmitter in the mammalian central nervous system. Thus reduce the activity of neurons, preventing nerve cells from overheating and decreasing blood pressure.

In addition, few of the lactic acid bacteria produce exopolysaccharides (EPSs), long-chain polysaccharides possess branched, repeating units of sugars or sugar derivatives. In LAB, EPSs typically exist in the cell surface or slime. Mainly exopolysaccharide are distinguished into homopolysaccharides (HoPS) and hetero polysaccharides (HePS). Majorly *Leuconostoc mesenteroides*, *Leuconostoc citreus*, and *S. thermophilus* isolated from animal, vegetables and fermented food produce HePS (Pessione., 2012).

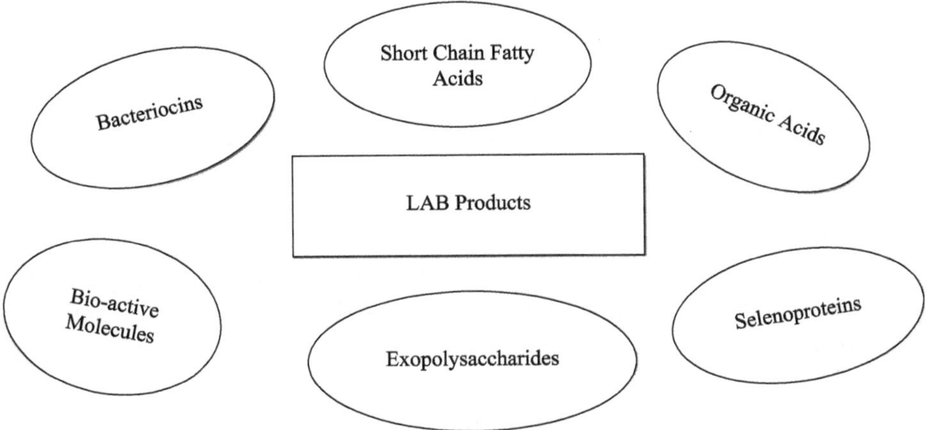

LAB host cross-talk and stress chaperons

From the recent studies it is known that in both bacteria and host "symbiosis" can trigger adaptive responses, that are predicted by transcriptomic and proteomic studies (host-microbiota interactomics), and focus on specific metabolic pathways activation, stress response and immune system stimulation/ depression.

The glycome in the cell wall of LAB plays a typical role in induction of cytokine in intestinal epithelial cells that has been proved with the use of mutants knock-out for protein glycosylation ability. These mutants not only failed to stimulate cytokines, but they were also unable to adhere to Caco-2 cells and showed a significant impaired persistence in the mice gut environment. The genetic loci in LAB specifically induced by gut transit are involved in nutritional adaptation (hydrolases, ABC transporters, and PTS systems) but also in anti-stress mechanisms like membrane modification, surface glycosylation and anti-oxidative defences. Bacterial stress protein production (enhanced synthesis of chaperones GroEL, GroES, DnaK, Hrc, Clp, Cts) in response to host factors has also been observed. In addition to this, it was well demonstrated that, besides helping correct protein folding, chaperones can perform multiple functions such as preventing denaturation, rearranging proteins after oxidative and pH stress and performing targeted proteolysis on the irreversible denaturated proteins. Out of these functions, is the adhesion to extracellular matrix proteins of the host tissues, and the stimulation of the host immune system play a key role in enhancement of immunity. To perform these additional functions chaperons should be surface or extracellularly located. This phenomenon is known as moonlighting and several moonlighting proteins have been reported both in eukaryotes and in bacteria, including LAB.

Effect of LAB on the Immune Response

Lactic acid bacteria play a significant role in various biological functions of the host. Many studies have demonstrated that LAB & milk fermented with LAB had antitumour activity and that they were able to prevent intestinal infection and demonstrated ole of LAB in immune function. *Lactobacillus plantarum* administered intraperitoneally stimulated the delayed type hypersensitivity (DTH) reaction, whereas non-viable bacteria acted as adjuvants for antibody production and *Lactobacillus casei* inoculated intraperitoneally activated peritoneal macrophages, increasing their phagocytic capacity.

On administration of *L.casei* Kupffer cells and the immune cells associated with spleen or lung and peritoneal macrophage were activated and also observed that the yogurt given to humans induced production of the cytokine interferon γ (IFN γ) when their circulating blood lymphocytes were stimulated with concavalin-A. On the contrary, there are few reports indicated that LAB do not always produce beneficial effects on the host, like *Lactobacillus rhamnosus* was associated with endocarditis or abscesses. Some strains of *Lactobacillus acidophilus* and *L. plantarum* may pose undesirable properties.

LAB adhesion to host tissues and immunomodulation

The bacteria-human mucosa interaction is at dynamic equilibrium and several factors can alter this balance. Intestinal movements and bacterial chemotaxis cause a weak association between bacteria and the surface of the gut epithelium where electrostatic and hydrophobic bonds are quite prominent. Specific interaction between molecular determinants on bacterial surface and specific gut mucosa receptors takes place during this course. Generally, specific targets on human tissues are collagen type I and IV, fibronectin, and laminin. In LAB, good binding-effectors are PS, lipoteichoic acids as well as proteins such as lectins and adhesins. The adhesion potential of a given LAB is very important for the survival and persistence in the gut environment.

Several surface located moonlighting proteins can act as adhesins. Trigger factor is also connected with adhesion: it has been described to be exposed on the surface of *Lactobacillus plantarum* and, in *L. reuteri* NCIB11951, a collagen I binding protein, shares high sequence homology with *E. coli* trigger factor. Apart from these proteins, glycolytic enzymes can have adhesive properties: GAPDH is able to bind fibronectin, plasmin, and mucine in several bacterial species and a similar role may be performed by PGK (phosphoglycerate kinase) that is referred to be extracellularly located in *Lactobacillus rhamnosusGG*.

IL-10 production is typically measured because it is an anti-inflammatory cytokine that suppresses IL-12 production and consequently IFN-production, thus favouring a T-helper 2 (Th2) or a regulatory T cell (Treg) response. In addition, IL-10 down-regulates antigen presentation and inhibits the activation of macrophages and thus the production of pro-inflammatory molecules and chemokines. IL-12 is also commonly measured as IL-12 production by DC is associated with the induction of T-helper 1 (Th1) responses. Further more IL-12 elicits IFN-γ production by T cells and NK cells. Studies have shown that the ability of different lactobacilli to induce a high ratio of IL-10/IL-12 or IL-10/TNF-α production in immune cells correlates with their capacity to provide significant protection in induced colitis in mice/rats models.

Key questions

1. The delayed type hypersensitivity (DTH) reaction is simulated on administration of:

   a) *Lactobacillus rhamnosus GG* b) *Lactobacillus plantarum*

   c) *Lactobacillus.reuteri* d) *Lactobacillus casei*

2. Neuroinhibitory molecule is:

   a) Ornithine b) GABA

   c) IFN γ d) Chaperons

3. The component of cell wall plays a crucial role in induction of cytokine in intestinal epithelial cells :

   a) Murine b) Glycome

   c) Peptdioglycan d) Suberin

4. The product of LAB used as biological preservatives in various food products in food industries:

   a) Exopolysaccharides b) Selenoproteins

   c) SCFAs d) Bacteriocin

Write a short notes on

i) Role of LAB in immune response.

ii) Short note on products of Lactic acid bacteria and their importance

iii) Importance of LAB in immunomodulation, discuss.

Further reading

Aleljung P, Shen W, Rozalska B, Hellman U, Ljungh A, Wadström T (1994) Purification of collagen-binding proteins of Lactobacillus reuteri NCIB 11951. CurrMicrobiol 28:231–236

Alvarez, R. A., Blaylock, M. W., & Baseman, J. B. (2003). Surface localized glyceraldehyde-3-phosphate dehydrogenase of Mycoplasma genitalium binds mucin. Molecular microbiology, 48(5), 1417-1425.

Juturu, V., & Wu, J. C. (2018). Microbial production of bacteriocins: Latest research development and applications. Biotechnology advances, 36(8), 2187-2200.

Lebeer, S., Claes, I. J., Verhoeven, T. L., Vanderleyden, J., & De Keersmaecker, S. C. (2011). Exopolysaccharides of Lactobacillus rhamnosus GG form a protective shield against innate immune factors in the intestine. Microbial biotechnology, 4(3), 368-374.

Pessione, E. (2012). Lactic acid bacteria contribution to gut microbiota complexity: lights and shadows. Frontiers in cellular and infection microbiology, 2, 86.

Sanchez, B., Bressollier, P., &Urdaci, M. C. (2008). Exported proteins in probiotic bacteria: adhesion to intestinal surfaces, host immunomodulation and molecular cross-talking with the host. FEMS Immunology & Medical Microbiology, 54(1), 1-17.

Sharpe, M., Hill, R. and Lapage, S. (1973) Pathogenic lactobacilli. J. Med. Microbiol. 6, 281-286.

Simone, C. D., Bianchi Salvadori, B., Negri, R., Ferrazzi, M., Baldinelli, L., &Vesely, R. (1986). The adjuvant effect of yogurt on production of gamma-interferon by Con A-stimulated human peripheral blood lymphocytes.

Wells, J. M. (2011). Immunomodulatory mechanisms of lactobacilli. Microbial cell factories, 10(Suppl 1), S17.

Index

A

Absorption: 4, 20, 23, 24, 25, 55, 73, 74, 75, 77, 84, 99, 107

Acidic: 2, 10, 20, 52, 55, 75, 76, 78, 79

Acidophilus: 14, 15, 16, 35, 36, 38, 39, 40, 41, 57, 61, 65, 70, 79, 86, 100, 107, 108, 109, 115, 121, 134

Acids: 2, 4, 5, 6, 17, 23, 25, 27, 29, 30, 35, 36, 38, 39, 56, 57, 60, 62, 65, 75, 76, 84, 92, 95, 102, 106, 108, 113, 114, 119, 123, 132, 133, 134

Action: 2, 3, 6, 9, 16, 17, 18, 49, 50, 56, 57, 61, 62, 74, 77, 81, 83, 99, 102, 104, 105, 107, 108, 114, 117, 118, 124, 125, 126, 129, 132

Activation: 3, 60, 61, 78, 81, 103, 133, 135

Addition: 3, 4, 5, 6, 8, 9, 11, 12, 15, 16, 23, 24, 25, 39, 40, 42, 52, 57, 58, 59, 60, 66, 67, 69, 74, 75, 76, 79, 82, 96, 100, 102, 103, 104, 108, 109, 110, 114, 127, 133, 135

Adhesion: 2, 4, 6, 9, 25, 54, 55, 56, 57, 63, 64, 70, 71, 74, 75, 80, 134, 136

Amino: 2, 4, 5, 10, 15, 17, 27, 29, 37, 38, 39, 52, 53, 67, 75, 102, 106, 114, 119, 121, 123, 124, 125, 128, 132

Amplified: 8, 35, 37, 42, 94, 95, 96, 97

Anaerobic: 15, 20, 38, 52, 54, 55, 63, 78, 96, 106, 108

Animal: 1, 7, 10, 11, 23, 43, 53, 54, 55, 57, 80, 87, 99, 100, 101, 104, 105, 106, 107, 108, 110, 111, 133

Antibiotic: 11, 14, 15, 16, 20, 53, 73, 74, 76, 79, 80, 87, 100, 101, 102, 104, 105, 114, 116, 126, 127

Antimicrobial: 2, 9, 18, 57, 58, 60, 62, 73, 74, 77, 81, 100, 101, 107, 113, 114, 116, 121, 123, 124, 125, 128, 129, 130, 132

B

Bacillus: 3, 14, 42, 76, 84, 91, 102, 103, 104, 107, 108, 111, 123, 124

Bacteria: 1, 2, 3, 4, 5, 6, 8, 9, 10, 12, 13, 14, 15, 16, 17, 19, 20, 21, 22, 23, 24, 33, 34, 35, 37, 38, 49, 50, 52, 53, 54, 55, 56, 57, 60, 61, 62, 63, 64, 65, 66, 67, 71, 73, 74, 75, 76, 77, 78, 81, 84, 89, 90, 91, 93, 95, 96, 97, 98, 99, 100, 101, 102, 103, 104, 105, 106, 107, 108, 109, 110, 111, 113, 114, 115, 116, 117, 118, 119, 120, 121, 123, 124, 125, 127, 129, 130, 131, 132, 133, 134, 135, 136

Bacteriocin: 3, 9, 60, 61, 62, 74, 113, 114, 116, 117, 118, 119, 120, 121, 123, 124, 127, 128, 130, 132, 135

Bacteriocins: 3, 6, 53, 56, 60, 61, 66, 75, 76, 78, 81, 87, 107, 113, 114, 116, 117, 118, 119, 121, 122, 125, 127, 129, 130, 132, 133, 136

Barrier: 3, 5, 6, 23, 56, 62, 64, 81, 102, 131

Benefit: 1, 11, 19, 23, 24, 26, 28, 31, 93, 102, 108

Benefits: 9, 13, 16, 18, 19, 22, 23, 25, 26, 28, 31, 33, 50, 62, 78, 99, 104, 107, 109, 110, 131, 132

Bifidobacteria: 15, 18, 20, 21, 22, 23, 24, 25, 31, 33, 35, 36, 37, 39, 41, 42, 43, 44, 45, 46, 47, 50, 55, 66, 74, 76, 81, 83, 84, 106, 107

Bifidobacterium: 5, 13, 15, 37, 42, 44, 45, 46, 47, 79, 104, 107, 113, 127

Biochemical: 4, 34, 35, 37, 38, 42, 45, 53, 114, 118, 132

Biological: 4, 12, 34, 45, 62, 71, 102, 103, 106, 129, 130, 132, 134, 135

Biology: 37, 46, 53, 55, 64, 95, 129, 130

Biosynthesis: 53, 63, 66, 68, 69, 70, 71, 116, 119, 124, 130

Blood: 15, 25, 44, 66, 109, 132, 134, 136

Bowel: 14, 15, 16, 24, 25, 55, 76, 80, 84, 89, 131

Bulgaricus: 39, 41, 59, 65, 69, 82, 86, 107, 108, 115, 121

Butyrate: 2, 3, 5, 23, 78, 84, 106, 132

C

Cancer: 6, 23, 24, 25, 30, 76, 77, 78, 87, 124, 127
Capable: 34, 38, 39, 51, 52, 76, 82, 85, 90, 96, 99
Carbohydrate: 21, 23, 42, 50, 51, 52, 53, 68, 71, 74, 84, 107, 113, 132
Carbohydrates: 4, 20, 26, 29, 51, 52, 56, 57, 113
Casei: 14, 15, 35, 36, 39, 40, 41, 46, 59, 60, 65, 69, 74, 83, 86, 115, 134, 135
Cells: 1, 2, 3, 9, 12, 19, 21, 24, 25, 50, 54, 55, 56, 57, 58, 59, 61, 62, 63, 64, 65, 71, 73, 74, 75, 76, 78, 80, 81, 84, 85, 86, 89, 108, 113, 116, 117, 119, 124, 128, 132, 133, 134, 135
Characteristics: 8, 9, 11, 15, 19, 20, 21, 34, 35, 37, 38, 50, 64, 65, 66, 89, 104, 114, 118
Characterization: 9, 42, 45, 53, 54, 55, 56, 57, 63, 78, 89, 91, 92, 98, 129, 130
Characterized: 1, 6, 10, 21, 57, 60, 67, 78, 85, 90, 113, 120
Cheese: 13, 15, 28, 40, 109, 115, 124, 126, 128, 129
Cholesterol: 5, 6, 23, 25, 30, 62, 66, 109
Classification: 33, 34, 37, 38, 40, 42, 45, 46, 49, 92, 118, 130
Clinical: 7, 8, 11, 12, 13, 16, 28, 31, 46, 81, 83, 84, 116, 127, 129, 132
Clostridium: 14, 16, 73, 74, 77, 84, 87, 106, 123, 124
Colon: 6, 15, 22, 23, 24, 25, 54, 56, 62, 76, 77, 80, 84, 87
Colonization: 2, 3, 16, 25, 53, 56, 62, 66, 81, 102, 105, 110
Comparative: 35, 38, 40, 46, 53, 58, 78, 87, 91, 97, 129
Complex: 4, 5, 6, 24, 36, 47, 49, 53, 54, 55, 58, 62, 90, 92, 94, 95, 96, 97, 117, 120, 125, 129, 131
Concentration: 2, 3, 12, 21, 60, 61, 84, 103, 107, 113
Conditions: 2, 3, 6, 9, 10, 11, 12, 15, 20, 23, 24, 55, 66, 68, 86, 91, 95, 106, 107, 108, 113, 119, 125
Confers: 2, 5, 19, 50, 75, 77, 79, 110
Consumption: 2, 10, 16, 24, 81, 82, 85, 110, 111
Containing: 15, 23, 24, 27, 28, 67, 85, 99, 103, 108, 111
Contrast: 7, 16, 26, 39, 52, 59, 60, 94, 95
Contribute: 1, 9, 40, 53, 61, 63, 66, 70, 89, 106, 110
Criteria: 7, 8, 11, 17, 20, 21, 27, 34, 49, 111
Culture: 1, 9, 20, 21, 22, 55, 89, 90, 91, 94, 96, 102, 103, 104, 109, 111, 115
Cytokines: 3, 58, 59, 60, 77, 81, 82, 83, 86, 87, 133

D

Dairy: 4, 13, 15, 28, 40, 52, 106, 109, 111, 115, 124, 125, 126, 127, 128, 129, 130
Decreased: 23, 24, 25, 55, 74, 83, 84, 85, 106, 107
Defined: 1, 19, 21, 26, 27, 34, 54, 55, 95, 96, 126
Derived: 4, 27, 54, 55, 77, 91, 99, 101, 106, 107, 109, 126
Diarrhea: 1, 6, 14, 16, 73, 74, 75, 76, 84, 85, 87, 109
Dietary: 4, 19, 26, 27, 28, 30, 54, 73, 104, 106, 111, 132
Digestion: 2, 4, 5, 9, 22, 54, 55, 56, 99, 101, 103, 104, 105, 106
Disease: 12, 16, 19, 23, 24, 26, 31, 76, 80, 81, 82, 83, 84, 85, 86, 87, 92, 100, 101, 104, 127, 131, 132
Disorders: 6, 7, 14, 24, 73, 80, 109, 127, 132
Diversity: 21, 38, 45, 47, 49, 53, 89, 92, 93, 95, 96, 97, 105, 113, 115, 117, 119, 121, 123, 125, 127, 129, 130
Drugs: 7, 11, 19, 30, 101, 108, 111, 126

E

Earlier: 7, 33, 35, 36, 37, 42, 54, 56, 59, 81, 99, 108, 116
Ecology: 33, 49, 51, 53, 55, 57, 59, 61, 62, 63, 90, 92, 127
Ecosystem: 49, 53, 54, 90, 94, 95, 116, 129
Electrophoresis: 8, 21, 35, 36, 42, 94, 95, 96, 97
Eliminating: 24, 54, 77, 102, 108, 109, 119, 124
Enterococcus: 15, 78, 84, 104, 107, 113, 114, 124, 127
Enzyme: 4, 10, 15, 35, 61, 68, 70, 75, 84, 85, 86, 94, 104
Epithelial: 2, 3, 5, 6, 9, 23, 50, 54, 55, 56, 57, 58, 59, 62, 63, 64, 73, 74, 75, 81, 82, 85, 133, 135

Epithelium: 6, 23, 56, 57, 58, 64, 65, 81, 85, 131, 134
Essential: 3, 5, 8, 40, 49, 50, 53, 61, 81, 82, 85, 92, 94, 98, 102, 105, 106, 118, 132
Exopolysaccharides: 65, 66, 68, 71, 87, 130, 133, 135, 136
Explain: 6, 9, 30, 45, 71, 74, 75, 80, 129
Expression: 3, 5, 10, 24, 56, 57, 58, 59, 69, 75, 76, 78, 83, 91, 92, 114, 117, 119
Extensively: 7, 15, 16, 76, 86, 101, 110, 125

F
Factor: 2, 3, 50, 54, 58, 59, 60, 86, 114, 134, 135
Factors: 2, 3, 7, 12, 23, 40, 49, 50, 53, 55, 56, 57, 60, 69, 73, 81, 85, 91, 100, 104, 110, 113, 117, 119, 124, 126, 133, 134, 136
Fatty: 4, 5, 23, 25, 28, 29, 30, 35, 60, 62, 74, 76, 84, 102, 106, 132, 133
Feces: 16, 20, 21, 25, 42, 43, 44, 77, 89, 105, 108
Ferment: 4, 19, 22, 34, 38, 51, 52, 113, 132
Fermentation: 8, 13, 15, 20, 21, 24, 25, 38, 42, 45, 51, 56, 62, 69, 78, 86, 106, 107, 109, 111, 113, 114, 116, 119, 124, 131, 132
Field: 7, 8, 11, 33, 35, 42, 54, 96, 101, 111, 129
Foods: 4, 7, 8, 12, 13, 15, 16, 23, 26, 27, 28, 30, 31, 35, 52, 63, 71, 74, 85, 86, 99, 100, 103, 108, 109, 110, 111, 113, 115, 117, 119, 121, 123, 124, 125, 127, 128, 129, 131
Formation: 3, 5, 6, 61, 68, 70, 77, 78, 82, 102, 107, 109, 117, 118, 125
Forms: 27, 28, 56, 57, 66, 95, 108, 117
Function: 3, 5, 6, 13, 24, 30, 37, 49, 54, 56, 61, 62, 64, 69, 81, 91, 117, 131, 134
Functional: 9, 13, 18, 19, 26, 27, 28, 31, 49, 50, 51, 53, 74, 78, 95, 97, 98, 99, 100, 126, 129, 131
Functions: 23, 26, 53, 54, 56, 58, 59, 71, 73, 76, 89, 92, 105, 117, 132, 133, 134

G
Gasseri: 15, 38, 39, 41, 52, 63, 79, 121
Gastric: 4, 6, 9, 10, 17, 20, 24, 30, 51, 54, 57, 66, 75, 85, 86, 99, 107
Gastrointestinal: 10, 14, 15, 16, 17, 19, 20, 21, 22, 23, 40, 49, 51, 53, 54, 55, 57, 59, 61, 62, 63, 70, 71, 74, 80, 82, 85, 87, 94, 127
Genera: 5, 19, 20, 33, 38, 40, 46, 50, 55, 102, 107, 109, 113
Genes: 3, 11, 13, 34, 36, 37, 38, 39, 40, 41, 44, 45, 46, 50, 51, 52, 53, 57, 58, 62, 63, 69, 75, 77, 78, 89, 91, 92, 93, 94, 98, 100, 101, 113, 116, 117, 119
Genetic: 8, 9, 24, 36, 40, 42, 52, 53, 62, 77, 85, 86, 91, 114, 116, 118, 119, 124, 132, 133
Genome: 36, 38, 39, 40, 45, 46, 47, 50, 51, 52, 53, 63, 78, 89, 90, 91, 92, 93, 95, 96, 97, 98, 113
Glucose: 8, 22, 39, 67, 68, 69, 70, 75, 84

H
Harmful: 1, 2, 7, 11, 19, 20, 56, 66, 74
Health: 1, 2, 7, 8, 9, 14, 15, 16, 17, 18, 19, 20, 22, 23, 25, 26, 27, 28, 30, 46, 49, 51, 62, 63, 66, 71, 73, 76, 78, 89, 92, 98, 99, 100, 101, 103, 104, 105, 108, 109, 110, 111, 114, 123, 131, 132
Healthy: 1, 3, 15, 22, 23, 26, 30, 49, 78, 81, 82, 92, 100, 105
Helveticus: 15, 39, 40, 52, 65, 70, 115, 121, 130
Human: 2, 4, 5, 7, 9, 10, 11, 12, 13, 15, 16, 18, 19, 20, 22, 23, 24, 30, 33, 39, 42, 43, 44, 46, 49, 50, 51, 52, 54, 57, 62, 63, 69, 70, 75, 78, 80, 85, 89, 91, 92, 93, 96, 98, 99, 100, 101, 105, 108, 110, 111, 114, 116, 124, 131, 132, 134, 136
Hybridization: 8, 21, 35, 36, 38, 42, 95, 97
Hydrogen: 21, 24, 56, 60, 78, 79, 81, 100, 110

I
Immune: 2, 3, 4, 9, 10, 11, 14, 23, 24, 55, 56, 57, 58, 59, 61, 64, 73, 75, 76, 77, 81, 83, 85, 86, 104, 105, 107, 108, 110, 114, 117, 124, 131, 133, 134, 135, 136
Immunity: 2, 3, 56, 57, 58, 59, 74, 79, 108, 116, 130, 131, 134
Industries: 18, 99, 100, 113, 114, 119, 132, 135
Infection: 2, 3, 12, 57, 58, 63, 64, 74, 76, 77, 79, 80, 87, 105, 127, 134, 136
Infections: 2, 3, 10, 11, 14, 23, 73, 74, 75, 76, 77, 78, 79, 87, 102, 104, 109, 126
Inflammation: 5, 23, 24, 25, 59, 76, 80, 81, 82, 83, 86, 132

Inflammatory: 3, 5, 6, 17, 23, 24, 25, 58, 59, 75, 76, 77, 80, 81, 82, 83, 102, 127, 131, 135
Inhibit: 59, 60, 75, 76, 77, 78, 80, 81, 118, 120, 123
Inulin: 21, 22, 23, 24, 30, 67, 69, 100
Isolate: 7, 8, 10, 33, 34, 37, 39, 45, 54, 55, 79, 90

J

Johnsonii: 14, 15, 38, 39, 41, 56, 61, 65, 67, 70
Journal: 18, 31, 46, 47, 63, 71, 111, 129, 130

K

Kefir: 13, 16, 28, 100, 108, 109, 115, 131

L

Lactate: 11, 23, 25, 51, 52, 84, 106, 107, 108
Lactic: 5, 6, 13, 14, 15, 17, 23, 38, 50, 52, 53, 60, 62, 63, 64, 65, 71, 75, 76, 81, 102, 107, 109, 110, 113, 115, 116, 117, 119, 121, 123, 125, 127, 129, 130, 131, 132, 133, 134, 135, 136
Lactobacilli: 15, 18, 20, 22, 25, 33, 35, 37, 38, 39, 41, 43, 45, 46, 47, 49, 50, 51, 52, 53, 54, 55, 56, 57, 58, 59, 61, 62, 63, 65, 66, 67, 69, 70, 71, 74, 75, 77, 78, 79, 80, 81, 84, 86, 107, 119, 121, 135, 136
Lactobacillus: 5, 13, 14, 15, 17, 18, 22, 24, 35, 38, 39, 40, 42, 45, 46, 47, 49, 50, 51, 52, 53, 56, 59, 60, 61, 62, 63, 64, 67, 70, 71, 74, 76, 79, 84, 85, 91, 102, 104, 106, 107, 108, 109, 115, 120, 121, 127, 130, 134, 135, 136
Lactococcus: 15, 16, 61, 91, 109, 113, 114, 115, 123, 132
Lactose: 4, 5, 6, 15, 22, 38, 69, 84, 85, 87, 109

M

Mediated: 5, 23, 59, 73, 75, 80, 85, 86, 132
Medical: 8, 10, 16, 26, 28, 63, 87, 108, 124, 136
Medicine: 11, 16, 25, 26, 30, 31, 63, 73, 99, 100, 111
Membrane: 3, 4, 10, 61, 65, 66, 68, 69, 85, 113, 117, 118, 125, 132, 133
Metabolic: 4, 5, 10, 11, 25, 38, 39, 45, 50, 51, 52, 53, 59, 62, 63, 68, 73, 75, 77, 78, 85, 89, 91, 93, 96, 106, 107, 108, 129, 131, 133
Metabolism: 5, 19, 23, 38, 39, 51, 52, 62, 63, 64, 68, 74, 77, 82, 85, 93, 104, 105, 107, 130, 132
Metabolites: 2, 4, 5, 6, 15, 25, 27, 61, 62, 64, 75, 77, 78, 84, 87, 108, 114, 131, 132
Microbes: 1, 18, 59, 60, 89, 90, 91, 98, 106, 127, 129, 131
Microbial: 1, 3, 4, 6, 7, 9, 16, 19, 20, 21, 23, 34, 49, 50, 53, 54, 55, 59, 61, 63, 64, 65, 71, 73, 74, 76, 78, 81, 86, 89, 90, 91, 92, 93, 94, 95, 96, 97, 99, 100, 101, 104, 105, 106, 110, 130, 136
Microflora: 1, 2, 19, 20, 22, 30, 49, 102, 105, 106, 107, 109
Microorganisms: 1, 2, 4, 5, 6, 10, 15, 16, 20, 21, 37, 49, 50, 52, 53, 54, 55, 63, 64, 74, 90, 92, 95, 97, 101, 105, 106, 108, 110, 111, 113, 125, 131
Modulate: 2, 3, 5, 23, 56, 58, 81, 127
Molecular: 8, 21, 22, 34, 35, 37, 46, 47, 53, 55, 57, 63, 66, 67, 68, 71, 73, 89, 90, 92, 95, 99, 114, 116, 117, 118, 119, 120, 121, 122, 123, 126, 130, 131, 132, 134, 136
Mucosal: 3, 16, 23, 67, 79, 81, 82, 83, 108

N

Natural: 7, 17, 26, 30, 49, 50, 76, 92, 93, 100, 104, 105, 113, 116
Nisin: 62, 66, 76, 113, 114, 115, 117, 118, 119, 121, 122, 123, 124, 125, 126, 127, 128, 129, 130, 132
Nutraceuticals: 19, 21, 23, 25, 26, 27, 29, 30, 31, 99, 111
Nutrient: 2, 4, 27, 38, 52, 55, 99, 107, 108, 113

O

Organic: 2, 6, 17, 34, 56, 62, 102, 103, 108, 111, 113, 132, 133
Origin: 33, 38, 43, 106, 111, 114, 129, 132

P

Pathogen: 2, 3, 9, 12, 65, 75, 76, 81, 83, 103, 119, 127, 131, 132
Pathogenic: 2, 6, 9, 16, 23, 54, 57, 60, 62, 63, 64, 84, 105, 109, 110, 114, 127, 136
Patients: 1, 10, 74, 79, 80, 81, 82, 83, 84, 86
Peptide: 3, 38, 60, 61, 62, 67, 84, 116, 119, 120, 121, 125, 130
Performance: 14, 53, 101, 104, 105, 107, 108, 127

Phylogenetic: 33, 34, 35, 36, 37, 38, 39, 40, 41, 42, 44, 45, 46, 91, 92, 93, 94, 95, 96, 97, 118
Physiological: 11, 12, 13, 23, 26, 30, 37, 40, 42, 45, 54, 75, 124, 132
Plantarum: 14, 15, 17, 38, 39, 41, 50, 52, 61, 63, 64, 65, 70, 81, 83, 84, 86, 91, 115, 132, 134, 135
Population: 6, 8, 20, 21, 30, 49, 53, 56, 61, 89, 91, 97, 107
Potential: 3, 4, 9, 11, 18, 23, 25, 27, 28, 30, 31, 49, 55, 61, 74, 75, 78, 81, 86, 93, 97, 104, 106, 108, 110, 111, 116, 117, 124, 126, 127, 128, 132, 134
Potentially: 9, 20, 25, 60, 63, 75, 82, 96
Prebiotic: 19, 20, 21, 22, 24, 28, 30, 66, 78, 80
Pressure: 25, 49, 50, 51, 84, 100, 109, 132
Primary: 5, 50, 56, 61, 62, 69, 79, 99, 102, 103, 114, 123
Probiotic: 1, 2, 3, 5, 6, 7, 8, 9, 10, 11, 12, 13, 14, 15, 16, 17, 18, 19, 22, 28, 31, 33, 34, 35, 37, 38, 49, 50, 53, 56, 57, 58, 59, 60, 61, 62, 63, 64, 65, 66, 74, 75, 76, 77, 78, 79, 80, 81, 82, 86, 87, 99, 100, 101, 102, 103, 104, 105, 107, 108, 109, 110, 111, 131, 136
Product: 4, 5, 11, 12, 13, 14, 19, 26, 27, 85, 91, 101, 128, 135
Production: 2, 3, 5, 6, 9, 11, 15, 16, 23, 24, 33, 34, 39, 53, 59, 60, 63, 65, 67, 68, 69, 70, 71, 73, 75, 76, 77, 78, 79, 81, 82, 83, 84, 91, 101, 104, 107, 110, 111, 113, 114, 115, 116, 119, 124, 126, 130, 133, 134, 135, 136
Promote: 3, 5, 22, 33, 56, 80, 103, 109, 110
Protein: 4, 5, 6, 16, 29, 35, 57, 58, 61, 66, 67, 85, 91, 106, 117, 122, 128, 130, 132, 133, 135

Q

Quality: 78, 80, 97, 100, 102, 103, 104, 105, 113

R

Rapid: 13, 33, 54, 75, 94, 97, 105, 106, 107, 128
Region: 9, 35, 38, 62, 67, 75, 79, 120
Regulation: 8, 10, 25, 56, 58, 64, 69, 75
Reported: 16, 23, 24, 39, 56, 57, 76, 103, 116, 121, 126, 134
Research: 1, 2, 6, 7, 8, 9, 11, 13, 16, 18, 21, 22, 30, 31, 53, 54, 55, 62, 73, 75, 87, 91, 95, 107, 116, 127, 129, 130, 136
Resulting: 3, 25, 50, 61, 75, 77, 78, 86, 91, 94, 95, 97, 107
Rhamnosus: 13, 15, 36, 40, 53, 65, 79, 80, 83, 84, 115, 134, 135, 136

S

Scfas: 4, 23, 24, 25, 62, 75, 76, 77, 78, 107, 132, 135
Scientific: 1, 7, 11, 12, 16, 26, 33, 99, 100, 111
Secreted: 16, 56, 58, 66, 116, 119, 127, 132
Secretion: 3, 55, 59, 68, 69, 75, 113, 117
Several: 2, 3, 5, 7, 8, 9, 10, 12, 13, 20, 27, 28, 39, 42, 54, 55, 57, 60, 65, 66, 68, 76, 78, 82, 86, 99, 102, 103, 104, 106, 108, 109, 119, 122, 126, 127, 132, 134, 135
Significant: 2, 7, 10, 11, 23, 30, 36, 39, 40, 50, 52, 58, 59, 62, 71, 74, 78, 93, 100, 101, 113, 127, 133, 134, 135
Species: 2, 3, 5, 6, 8, 11, 13, 14, 15, 16, 19, 20, 33, 34, 35, 36, 37, 38, 39, 40, 42, 43, 44, 45, 46, 47, 49, 50, 51, 52, 53, 55, 56, 59, 60, 61, 65, 67, 69, 70, 74, 77, 78, 79, 81, 82, 83, 84, 90, 92, 95, 96, 97, 102, 103, 104, 105, 107, 109, 114, 116, 118, 123, 126, 135
Staphylococcus: 61, 78, 84, 111, 123, 124, 125, 129
Strain: 6, 7, 8, 9, 11, 13, 16, 17, 33, 36, 39, 42, 60, 65, 83, 93, 96, 102, 111, 119, 126
Streptococcus: 14, 15, 61, 74, 85, 96, 104, 106, 107, 108, 113, 124, 125, 126
Sugar: 4, 20, 38, 39, 52, 66, 69, 133
Survival: 19, 24, 26, 39, 50, 75, 76, 102, 134
Synthesis: 5, 10, 23, 24, 25, 38, 39, 62, 67, 68, 69, 70, 71, 91, 93, 94, 106, 117, 125, 129, 133

T

Taxonomic: 33, 34, 36, 40, 42, 45, 46, 96
Taxonomy: 33, 34, 35, 37, 40, 45, 46, 47
Techniques: 8, 21, 35, 37, 40, 46, 55, 90, 92, 93, 94, 95, 96, 97, 111, 114, 126
Temperature: 12, 36, 37, 91, 94, 95, 102, 106, 114, 117, 119
Terminal: 55, 57, 61, 67, 94, 95, 116, 118, 123
Therapeutic: 7, 11, 13, 16, 62, 105, 108, 116, 127, 131, 132
Therapy: 1, 16, 31, 74, 75, 79, 80, 81, 86, 87, 101
Toxic: 9, 25, 61, 66, 102, 113, 114, 118

Transfer: 10, 11, 50, 52, 62, 63, 68, 69, 100, 105
Treat: 16, 76, 79, 81, 82, 108, 122, 126
Treatment: 12, 16, 24, 26, 74, 78, 80, 81, 82, 86, 87, 102, 109, 127, 129, 132

V

Variety: 11, 15, 45, 51, 55, 58, 60, 65, 89, 90, 91, 99, 100, 105, 111, 131
Various: 5, 7, 9, 25, 34, 53, 54, 59, 62, 67, 69, 73, 76, 85, 91, 103, 108, 113, 116, 126, 131, 132, 134, 135
Viable: 1, 9, 12, 19, 89, 98, 101, 116, 134
Vitamins: 4, 5, 15, 24, 26, 27, 73, 102, 106, 108, 109
Vitro: 2, 9, 10, 11, 13, 20, 22, 31, 53, 54, 57, 63, 110, 127, 129

W

Water: 9, 21, 67, 73, 100, 101, 102, 103, 108, 132
Weight: 85, 102, 104, 114, 116, 117, 118, 119, 121, 123, 132

Y

Yogurt: 13, 15, 16, 28, 43, 85, 108, 109, 131, 134, 136